ALGEBRA EXAMPLES

TRIGONOMETRY 3

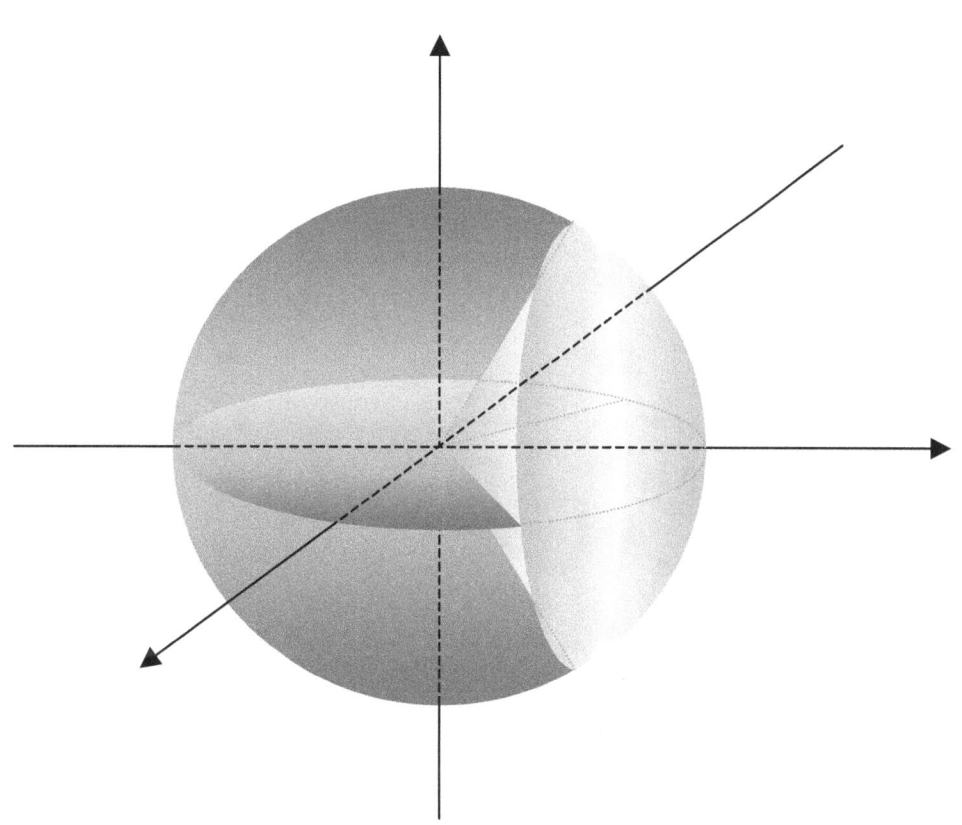

Seong R. KIM

Dear students:

Students need the best teacher, so you need examples, because examples are the best teacher. All the examples here are fully worked, and explain **how** the basic and essential tools in math are made, together with **what** they are, **how** they work, and **how** to work with them. Such tools include numbers, formulas, identities, equations, laws, etc.

Examples here begin with easy ones, of course. Covering every meter and yard properly, we can cover thousands of miles and kilometers. And it is particularly the case in math.

Of those examples therefore, some might even look too easy for you. It's not that easy though, to come up with those examples. Anyways, the bigger and the taller the tree, the deeper and the stronger the root.

Doing math, we work with ideas and run ideas, because every thing in math is an idea. A number is an idea, for instance, and the same is true for a line or circle, too. And putting ideas together, we build another, which becomes the base or an element of another, and each is connected. And that's the way your math grows. So you get to build a circuit, and sometimes, need to fill the gap or repair the circuit so that you get the sense of it.

So your calculation runs properly, and you get the problem solved.

The examples have been made and arranged so that they get tougher (or sometimes easier for some reason) as you proceed with them. In particular, similar examples with some variations are strategically repeated so that you can get the ideas or the tools tricky or complicated, and can get them mastered.

This book is however, nothing but a bunch of examples until you get it powered. How then, to get it powered, and make it run and work for you?

Just read it, and then, do each example in writing. And it is important to note that you do it in **your** writing. Just watching someone doing it, you just only feel that you can do it. If you do it, you can do it, but if you don't, we can hardly. It's a cliché, of course, but is always true that knowing is one thing and doing is another.

I've been helping students grow, take care of, and run their own math. The area covers algebra and geometry for high school or college students, and is especially for equations (for unknowns or curves), functions, and their graphs, which are the basic elements in calculus, which's been the core of my interest from my early age in high school.

Of my students, some are quite poor in math, and thus, are afraid of or hate math, some require special education because of exceptional intelligence, some are smart enough, some are naïve and diligent, some are clever but lazy, and most behave in general. All the students are badly after though, one thing in common: a strong and secure math skill. It is of course, the prime objective of my work, and I'm always happy to and eager to help them achieve it. The problem was however, that many of them wanted it to be purchased. And the question is, can we buy it?

We can buy the means, of course. And a solid math skill is feasible, too. We know however, we can't buy love, and the same is true for the math skill, too. It's not what we can buy or sell, and not what we can give or take. It is however, what we can grow, and need to grow. Your math grows as much as you grow and take care of it. So does mine.

What math then, do students most often do or use in high schools or colleges?

It is algebra and geometry. What algebra though?

Elementary algebra, of course
Doing the algebra, we work with numbers (many in kinds), constants, variables, ratios, rates, expressions, equations, inequalities, functions, identities, formulas, laws, etc., together with signs and symbols. And if we want to do algebra properly, we want to know their natures and how they mingle with each other.

So studying math ideas or tools, you want to know **what** they are, **how** they work, and **how** to work with them or **what** to do with them.   What then, about the geometry?

Basically, the geometry has much to do with shapes, positions, and angles. The shapes begin with triangles and circles, and move on to rectangles, squares, parallelograms or rhombuses, trapezoids, tetragons, other polygons, polyhedrons, etc.

Doing the geometry, too, though, we need to do the algebra stated above. So it is analytic geometry, often called coordinate geometry, too. And doing it, we can specify positions using coordinates. So in the geometry, basically, we work with graphs. Putting a math idea in a graph, we can not only effectively think about it but actually see it, too, and therefore, can efficiently work with it. What idea then, is it?

The idea begins with a point, line, parabola, circle, ellipse, and hyperbola, called a conic section or basic curve, and then, moves on to other curves, planes, surfaces, volumes, and other objects in various dimensional spaces, together with vectors.

And using an angle, we can specify an amount of turn or change in direction.

So learning, using, or applying those ideas or math tools, we get to solve problems.

And this book can help. It can help learn them, and use them so that you can navigate to find solutions to problems. And in particular, it can help come up with answers to those **what**s and **how**s stated above. So it can help you grow and run your own math, and thus, can help achieve your solid math skill.

It is however, not a magic book giving you a math skill of high caliber overnight. And it can have many mistakes, too. There is no magic, and math is full of facts and ideas. And it is after all, not me and not your teacher but you who put together some of those facts and ideas, and understand it. Putting facts and ideas together, understanding it, and taking care of what you have learned, you grow your math. And this book can help.

This is a book of examples designed to help you grow your math, and assumes that you are a real beginner. This book requires though, time and effort, the amount of which need to be substantial, too, but will be worth it. That's because you want a substantial achievement, and will get it. And probably, you will get to see this book helping you get there much faster than expected. And then, you will get to see the way math runs.

In math, everything is an idea. So is a problem. And solving it, we put it many different ways. For instance, while expanding or reducing it, or modifying or converting it, we keep searching for the solution, approaching the solution, and eventually, can get there. So don't look for the solution outside the problem. The solution is inside the problem if the problem is properly made.

If it is not, no solution is the solution. And in fact, it is often the case a problem itself is the solution. We can put a problem in many different ways, and eventually, can end up with the solution.   How come then, is the solution no other than the problem?

For instance, the solution to $3232 \div 101$ is 32.   And we can put it this way:

$$3232 \div 101 = \frac{3232}{101} = \frac{32 \times 101}{101} = \frac{32}{1} = 32 \Rightarrow 3232 \div 101 = 32.$$

And we can get this, too: $32 \Rightarrow 3232 \div 101$.   How?

$$32 = \frac{32}{1} = \frac{32 \times 101}{101} = \frac{3232}{101} = 3232/101 = 3232 \div 101.   \text{ Too easy?}$$

For another instance, the solution to $ax^2 + bx + c = 0$ is: $x = \frac{-b \pm \sqrt{b^2 - 4ac}}{2a}$, which is called the quadratic formula.   How come then, is the solution no other than the problem?

We can put it this way:

$$x = \frac{-b \pm \sqrt{b^2-4ac}}{2a} \Rightarrow 2ax = -b \pm \sqrt{b^2 - 4ac} \Rightarrow 2ax + b = \pm \sqrt{b^2 - 4ac}$$

$$\Rightarrow (2ax + b)^2 = b^2 - 4ac \Rightarrow 4a^2x^2 + 4abx + b^2 = b^2 - 4ac$$

$$\Rightarrow 4a^2x^2 + 4abx = -4ac \Rightarrow ax^2 + bx = -c \Rightarrow ax^2 + bx + c = 0.$$

And we can get this, too: $ax^2 + bx + c = 0 \Rightarrow x = \frac{-b \pm \sqrt{b^2-4ac}}{2a}$. How?

$$ax^2 + bx + c = a(x^2 + \tfrac{b}{a} x) + c = a(x^2 + \tfrac{b}{a} x + \tfrac{b^2}{4a^2} - \tfrac{b^2}{4a^2}) + c = a(x^2 + \tfrac{b}{a} x + \tfrac{b^2}{4a^2}) - \tfrac{b^2}{4a} + c$$

$$= a(x + \tfrac{b}{2a})^2 - \tfrac{b^2-4ac}{4a} = 0 \Rightarrow a(x + \tfrac{b}{2a})^2 = \tfrac{b^2-4ac}{4a} \Rightarrow (x + \tfrac{b}{2a})^2 = \tfrac{b^2-4ac}{4a^2} \Rightarrow x + \tfrac{b}{2a} = \pm\sqrt{\tfrac{b^2-4ac}{4a^2}}$$

$$\Rightarrow x = -\tfrac{b}{2a} \pm \tfrac{\sqrt{b^2-4ac}}{2a} = \tfrac{-b \pm \sqrt{b^2-4ac}}{2a} \Rightarrow x = \tfrac{-b \pm \sqrt{b^2-4ac}}{2a}.$$

And we call the set of processes above, algebra.

So if a problem is well defined, that is, if it makes sense, we should be able to get it solved the way below:

A problem $\Rightarrow$ ... $\Rightarrow$ ... $\Rightarrow$ the solution, and thus: **the problem $\Rightarrow$ the solution**.

So solving a problem, we put it many different ways so that we can get to the solution.

And that's the way, math runs.

May your math run very well.

Seong R. Kim

B.S. Math. Michigan Tech. Univ.   M.S. Math. Rensselaer Polytechnic Institute

Notes:

This book is about a math idea called trigonometry.

Why trigonometry though?

That's primarily because we often get to work with angles not only doing geometry but doing algebra, too. Doing high school math or college math, we can hardly avoid or stay away from algebra. And doing geometry, too, we often get to do algebra on expressions with angles. And doing such algebra, we can say that we do trig-algebra.

And next, we need to do trigonometry if we have to work with vectors and many other objects that have to do with angles. Working with such objects, we often need to find objects called components, which have directions or angles. Finding such components, we want to use some tools in trigonometry. And the tools are called trigonometric-ratios, often just called trig-ratios, for short.

So doing trigonometry, we get to use trig-ratios called sines, cosines, etc., together with a bit bigger tools called trig-identities, and some rules or formulas. So in this book, you get to know what those tools are and what they are about. That is to say that you will get to know how those tools work, what you can do with them, and how to work with them.

Specifically, you will get to learn, for instance, how the trig-ratio called the sine is made, what it is about, and how to use it. More specifically, you will see why you have to multiply by the sine, and what you get multiplying by it.

And of course, you will get to learn those important tools called trig-identities and the formulas called the Sine Rule and the Cosine Rule. Besides, you will get to see and will be familiar with some special tools called trig-functions.

And you will get them all through examples, that is, those tools will get explained with examples fully worked and detailed. Also, following steps to the solution in each example, you will be more familiar with the tools and the math ideas.

And you will get to strengthen your skill of algebra. So doing problems as well as learning ideas in math, you can do better and faster so that your math can run not only properly but fast enough, too.

And all the basics, tools, and ideas are covered in three books as well as in one book. And the three are as follows:

ALGEBRA EXAMPLES TRIGONOMETRY 1, which covers from the section Intro 1 to the section The Cosine Rule.

ALGEBRA EXAMPLES TRIGONOMETRY 2, which covers from Examples 1 in The Cosine Rule to the section Sine Functions.

ALGEBRA EXAMPLES TRIGONOMETRY 3, which is this book, and covers from Examples in Sine Functions to Examples in Trig-Algebra.

And all the contents of the three books above are put in one book as follows:

ALGEBRA EXAMPLES TRIGONOMETRY, which covers thus, from the section Intro 1 to Examples in Trig-Algebra.

Contents

In TRIGONOMETRY 3

The Preview of the **Contents**

In TRIGONOMETRY 1

The Preview of the Contents

In TRIGONOMETRY 2

$$(x+y)^2 = x^2 + 2xy + y^2. \qquad\qquad (x+y)^3 = x^3 + 3x^2y + 3xy^2 + y^3.$$

$$(x+y)(x-y) = x^2 - y^2. \qquad\qquad (x+y)(x^2 - xy + y^2) = x^3 + y^3.$$

$$(x^2 + xy + y^2)(x^2 - xy + y^2) = x^4 + x^2y^2 + y^4.$$

$$(x+a)(x+b) = x^2 + (a+b)x + ab. \qquad (ax+b)(cx+d) = acx^2 + (ad+bc)x + bd.$$

$$(x+a)(x+b)(x+c) = x^3 + (a+b+c)x^2 + (ac+bc+ca)x + abc.$$

$$(a+b+c)^2 = a^2 + b^2 + c^2 + 2(ab+bc+ca).$$

$$(a+b+c)(a^2 + b^2 + c^2 - ab - bc - ca) = a^3 + b^3 + c^3 - 3abc.$$

Suppose both a and $b \neq 0$, and both m and n are integers. Then, we get:

0. $a^m a^n = a^{m+n}$

1. $a^m/a^n = \dfrac{a^m}{a^n} = a^{m-n}$

2. $(a^m)^n = a^{mn}$

3. $(ab)^n = a^n b^n$

4. $(a/b)^n = \left(\dfrac{a}{b}\right)^n = a^n/b^n = \dfrac{a^n}{b^n}$

Suppose both a and $b > 0$, and m and n both are integers nonzero. Then, we get:

0.1. $a^{\frac{1}{n}} b^{\frac{1}{n}} = (ab)^{\frac{1}{n}}$.

1.1. $\dfrac{a^{\frac{1}{n}}}{b^{\frac{1}{n}}} = \left(\dfrac{a}{b}\right)^{\frac{1}{n}}$.

2.1. $(a^{\frac{1}{n}})^m = (a^m)^{\frac{1}{n}}$.

3.1. $(a^{\frac{1}{n}})^{\frac{1}{m}} = a^{\frac{1}{mn}} = (a^{\frac{1}{m}})^{\frac{1}{n}}$.

3.2. $(a^{mp})^{\frac{1}{np}} = (a^m)^{\frac{1}{n}}$, where p is a nonzero integer.

1. Suppose M, N, and $b > 0$, but $b \neq 1$, and we have: $A = \log_b M$, and $B = \log_b N$.
Then, we get: $A - B = \log_b M - \log_b N = \log_b \frac{M}{N}$.

2. Suppose that M and $b > 0$, but $b \neq 1$, and that we have: $E = \log_b M$.
Then, we get: $PE = P \log_b M = \log_b M^P$.

3. Suppose that a, b, C, and $D > 0$, but a and $b \neq 1$, and that we have: $\log_a C = \log_b D$.
Then, we get: $\log_a C = \log_b D = \log_{ab} CD$.

4. Suppose that a, b, C, and $D > 0$, but a and $b \neq 1$, and that we have: $\log_a C = \log_b D$.
Then, we get: $\log_a C = \log_b D = \log_{\frac{a}{b}} \frac{C}{D} = \log_{\frac{b}{a}} \frac{D}{C}$.

5. $\log_b b = 1$, and $\log_b 1 = 0$.

6. $\log_b A = \dfrac{\log_c A}{\log_c b}$.

7. $\log_b A = \dfrac{1}{\log_A b}$.

Note:

The drawings or graphs in this book are not exact, and are approximate or conceptual ones.

$\in$	"$a \in B$" means that a belongs to B. "$p, q,$ and $r \in W$" means that $p, q,$ and r belong to W.
$\Rightarrow$	"$A \Rightarrow B$." means that A implies B.
$\equiv$	$A \equiv B$ means that A and B are identical to each other.
$\neq$	$A \neq B$ means that A is not equal to B.
$|A|$	The magnitude of A. For instance, $|-1| = |1| = 1$.
$\therefore$	Therefore
$\Leftrightarrow$	"$A \Leftrightarrow B$" means "If A then B." and "If B then A." We can read $A \Leftrightarrow B$ as "A if and only if B." In such a case, we can say that $A = B$.
Δx and Δy	Suppose that (x_1, y_1) and (x_2, y_2) are two points in the x-y plane. Then, we get either of the two below. $\Delta x = x_2 - x_1$, and $\Delta y = y_2 - y_1$. $\Delta x = x_1 - x_2$, and $\Delta y = y_1 - y_2$.

Distance Formula

Suppose that d is the distance between two points (x_1, y_1) and (x_2, y_2) in the x-y plane. Then, we get $d^2 = (\Delta x)^2 + (\Delta y)^2$.

Examples in Sine Functions

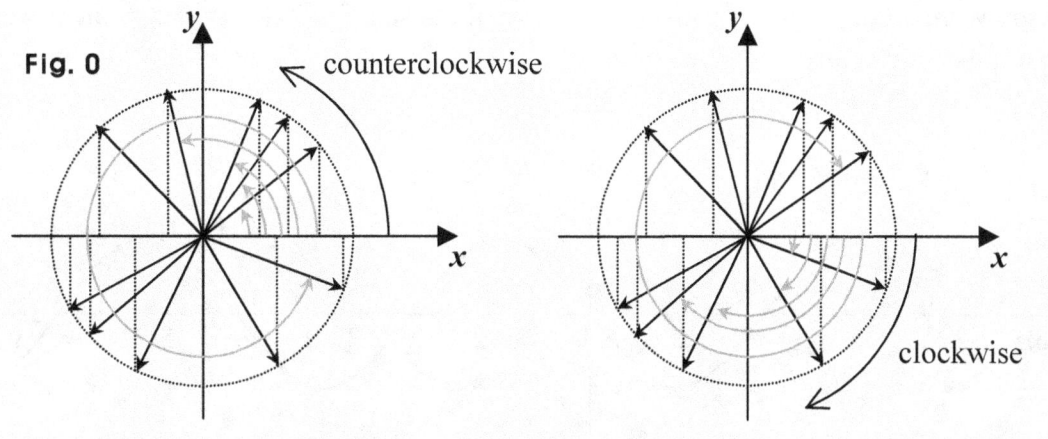

Fig. 0   counterclockwise   clockwise

In trigonometry dynamic, the ray turning makes governing angles. If turning clockwise, it makes angles negative, and if counterclockwise, it makes angles positive. And of course, if no turning, the angle made is 0.

So a governing angle can be $0°$ or any angle positive or negative.
In each right triangle, the ray is of length 1, and is the hypotenuse.
And (x, y) is the terminal point, so x is the adjacent, and y is the opposite.

And thus, assuming θ is a governing angle, we get: $\sin \theta = y$.
So if the ray is in the first quadrant, $\sin \theta > 0$, since $y > 0$.
In the second, $\sin \theta > 0$, too, since $y > 0$.
In the third, $\sin \theta < 0$, because $y < 0$. And in the fourth, $\sin \theta < 0$, too, since $y < 0$.

Put in a graph the curve of each of the equations below:

0. $y = \sin |2x|$

1. $|y| = \sin 2x$

2. $|y| = \sin |2x|$

3. $y = \sin x + \sin |x|$

Suggestions or Solutions
To the **Problem** in the Example **0**

Put in a graph the equation as follows: $y = \sin |2x|$.

To begin with, putting in the *x-y* plane, the curve of the sine function $y = f(x) = \sin x$, we can put it the way below:

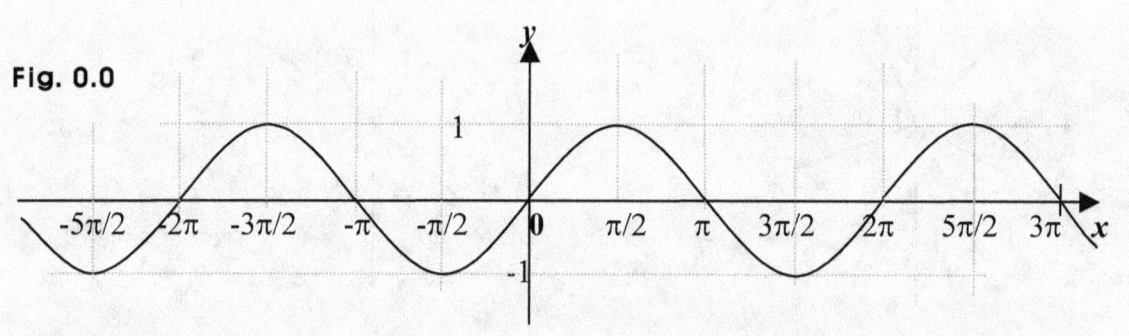

Fig. 0.0

The curve above is in fact, the same as the curve of an equation $y = \sin x$.

And the curve of an equation $y = \sin 2x$ is the curve of a function $y = g(x) = \sin 2x$, too.

We know in the curve of the function g, the frequency is 2, so the period is $2\pi/2 = \pi$.

So putting in a graph, the curve of the equation $y = \sin 2x$, we can put it this way:

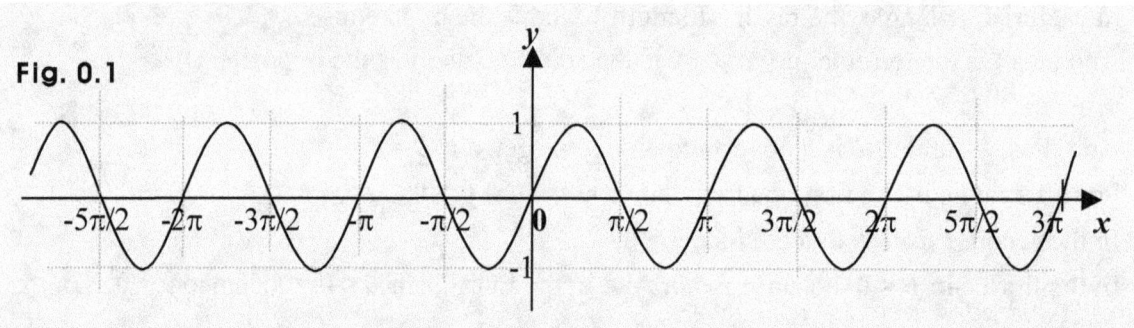

Fig. 0.1

What then, about the curve of $y = \sin |x|$?

In that case, we want to consider two cases, one is: $x \geq 0$, and the other is: $x < 0$. Why?

Though looking awkward, **sin |x|** can be put this way, too: **sin t** where *t* = |x|.

And we know: **|x| ≥ 0** for all *x*.

So even if *x* < 0, that is, *x* gets a negative value, we still get: *t* > 0, that is, *t* gets a positive value.

Thus, for instance, even if *x* = **-π/6**, we need to get: **sin π/6**, because |-π/6| = π/6.

So given *y* = **sin |x|** for *x* ≥ 0, we can just set: *y* = **sin x**.

If however, *x* < 0, we want to set: *y* = **sin (-x)**, because –*x* > 0, since |x| ≥ 0 in **sin |x|**.

And we have a trig-identity: **sin (-x) = -sin x**.

And thus, given *y* = **sin |x|**, we get: *y* = **sin x** for *x* ≥ 0, and *y* = **-sin x** for *x* < 0.

So taking care of first, the curve of *y* = **sin x** for *x* ≥ 0, we can put it the way below:

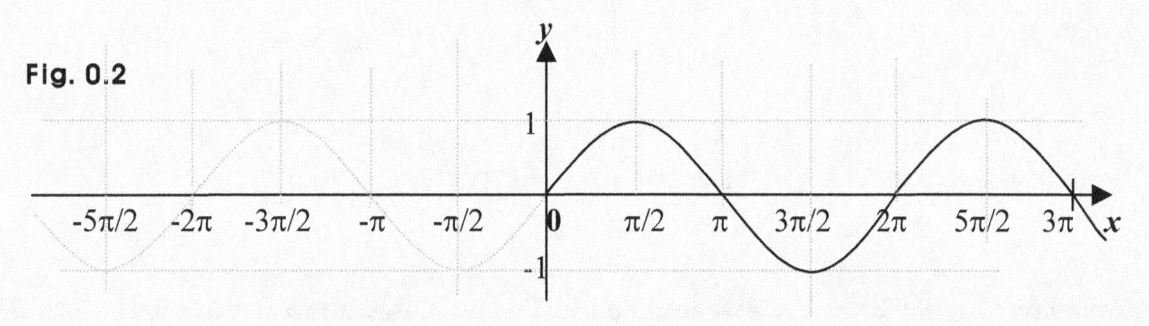

Fig. 0.2

Next, moving on to the curve of *y* = **-sin x**, we can put it the way below:

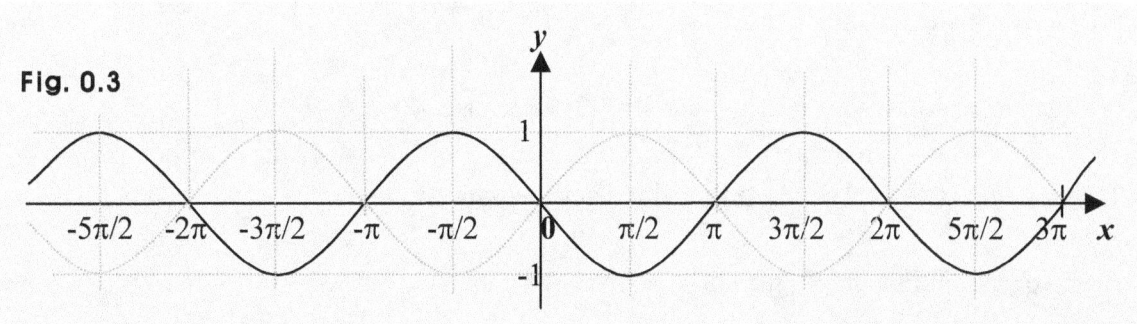

Fig. 0.3

So next, moving on to the curve of $y = -\sin x$ for $x < 0$, we can put it the way below:

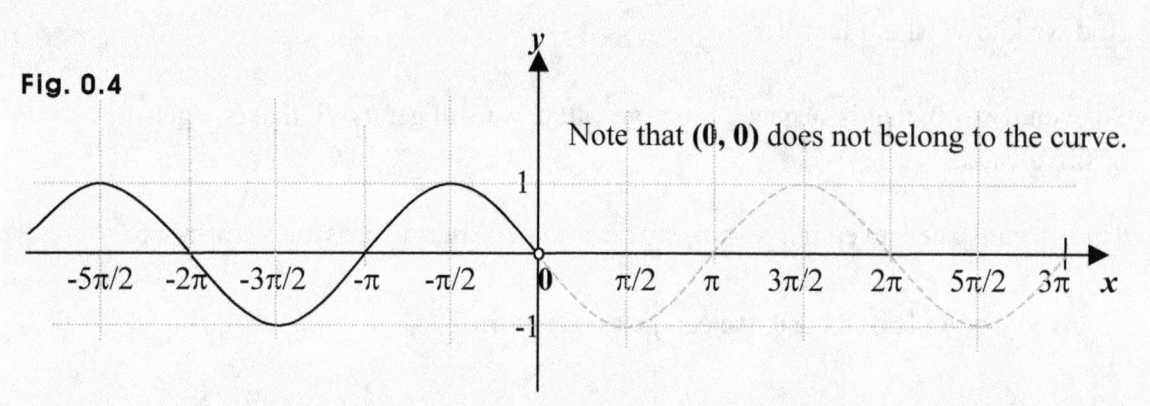

Fig. 0.4

Note that **(0, 0)** does not belong to the curve.

And thus, putting in a graph, the curve of $y = \sin |x|$, we can put it the way below:

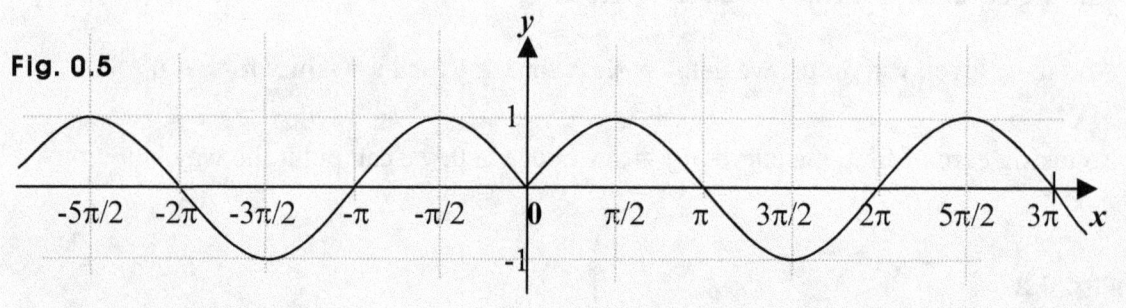

Fig. 0.5

What then, about the curve of the equation $y = \sin |2x|$?

We know: $|2x| \geq 0$ for all x. So we want to consider two cases: $x \geq 0$, and $x < 0$.

That is, $y = \sin |2x|$ is the same as $y = \sin 2|x|$.

So given $y = \sin 2|x|$, and if $x \geq 0$, we can set: $y = \sin 2x$.

If however, $x < 0$, we want to set: $y = \sin (-2x)$ because $-2x > 0$.

Thus, we get: $y = \sin 2x$ for $x \geq 0$, and we get: $y = -\sin 2x$ for $x < 0$.

And we can put in a graph, the curve of $y = \sin 2x$ the way below:

Fig. 0.6

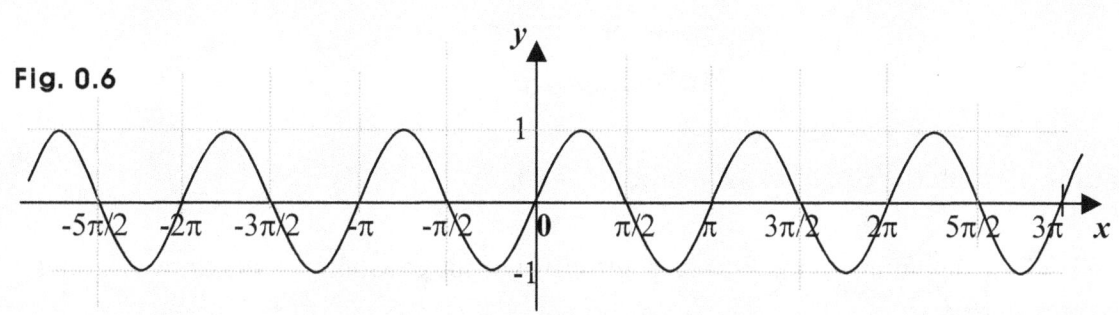

So we can put in a graph, the curve of the equation $y = \sin |2x|$ the way below:

Fig. 0.7

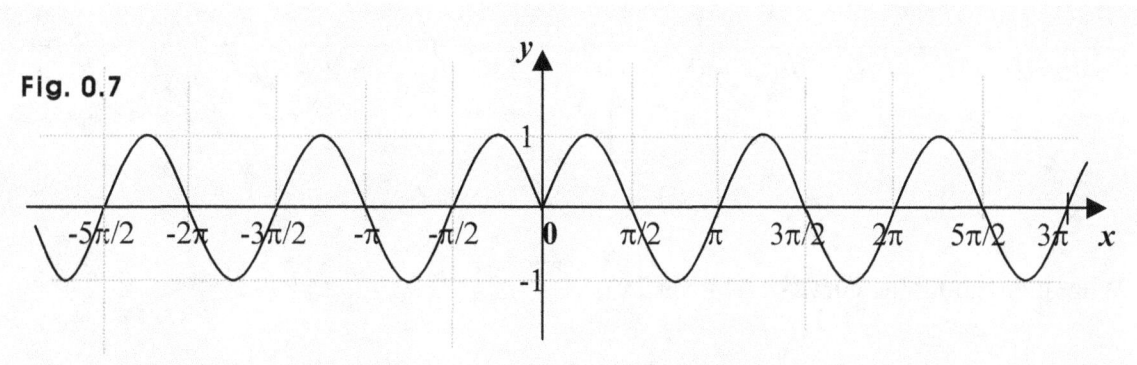

Notice that the curve is symmetric about the y-axis.

Suggestions or Solutions
To the **Problem** in the Example **1**

Put in a graph the equation as follows: $|y| = \sin 2x$.

To begin with, putting in the *x-y* plane, the curve of the equation $y = \sin 2x$, we can put it the way below:

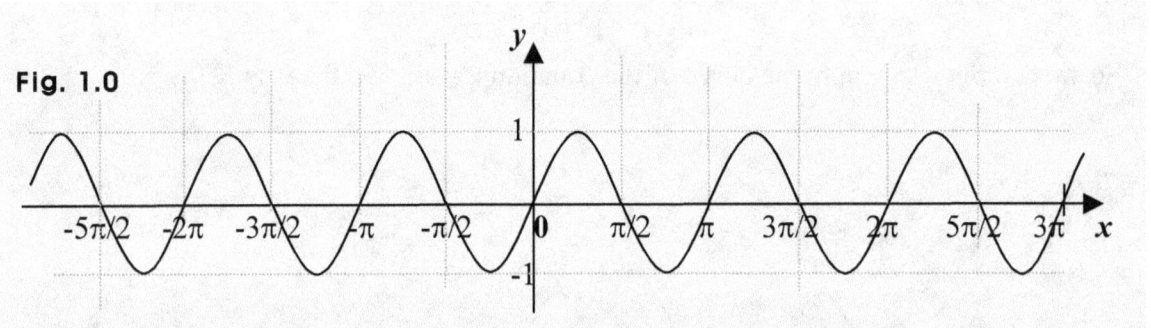

Fig. 1.0

What then, about the curve of $|y| = \sin 2x$?

In that case, we want to consider two cases, one is $y \geq 0$, and the other is $y < 0$.　Why?

Though looking awkward, $|y| = \sin 2x$ can be put this way, too: $t = \sin 2x$ where $t = |y|$.

And we know: $|y| \geq 0$ for all *y*.

So even if $y < 0$, that is, *y* gets a negative value, we still get: $t > 0$, that is, *t* gets a positive value.
So for instance, even if $y = -1/2$, we need to get: $t = 1/2$, because $|-1/2| = 1/2$.

So given $|y| = \sin 2x$ for $y \geq 0$, we can just set: $y = \sin 2x$.

If however, $y < 0$, we want to set: $-y = \sin 2x$ because $-y > 0$ since $|y| \geq 0$ in $|y| = \sin 2x$.

Thus, we get: $y = \sin 2x$ for $y \geq 0$, and we get: $y = -\sin 2x$ for $y < 0$.

And taking care of first, the curve of $y = \sin 2x$ for $y \geq 0$, we can put it this way:

Fig. 1.1

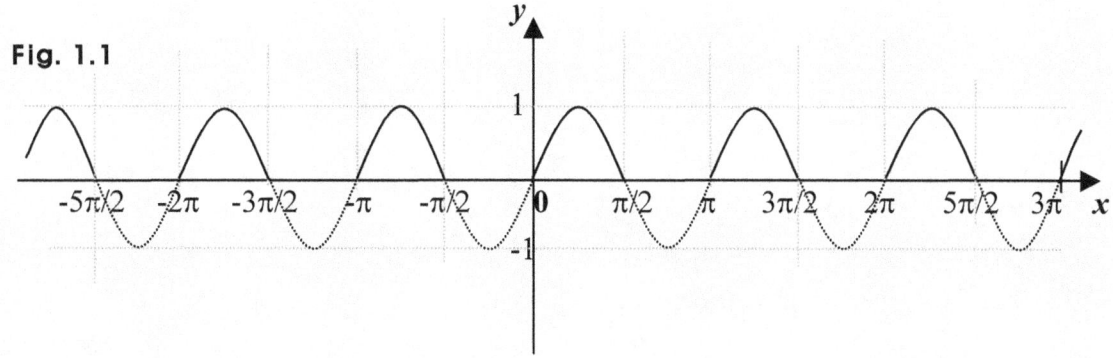

Next, we can put in a graph, the curve of $y = -\sin 2x$ the way below:

Fig. 1.2

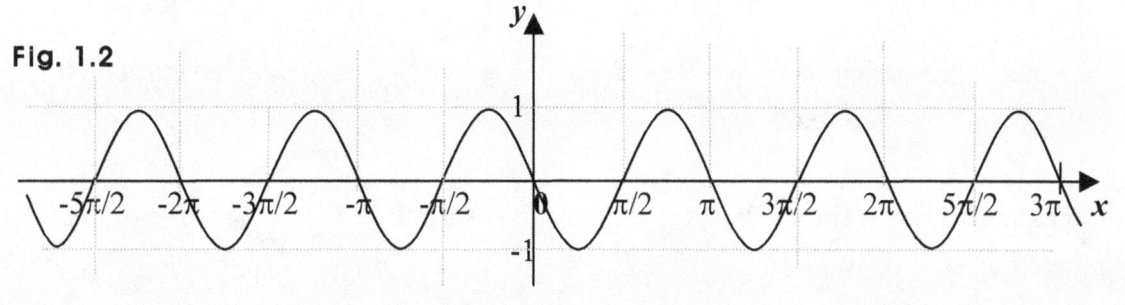

So next, moving on to the curve of $y = -\sin 2x$ for $y < 0$, we can put it this way:

Fig. 1.3

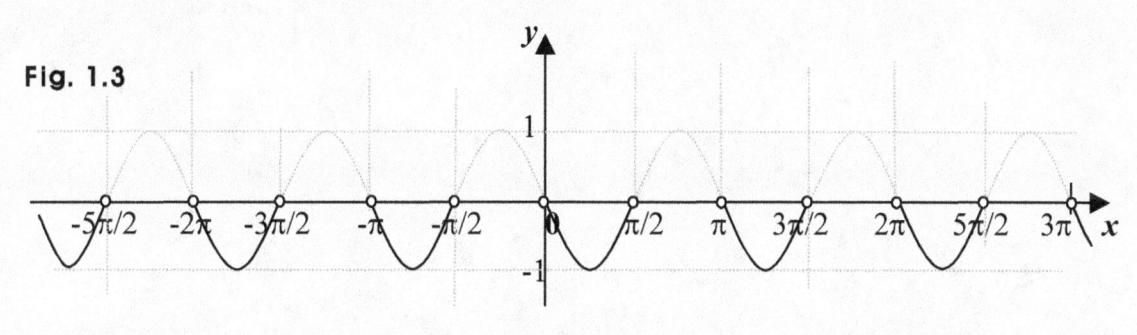

Note that all the points $(n\pi/2, 0)$ for n integer, do not belong to the curve above.

And thus, putting in a graph, the curve of $|y| = \sin 2x$, we can put it the way below:

Fig. 1.4

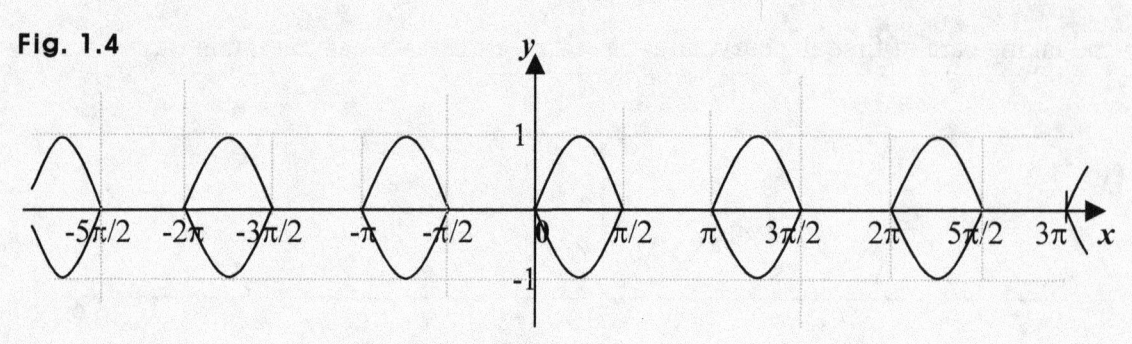

Notice that the curve is symmetric about the **x**-axis.

In fact, the curve of an equation |**y**| = **f**(**x**) is symmetric about the **x**-axis.

For instance, assuming $f(x) = x^2 - 3x + 2$, we get: $|y| = x^2 - 3x + 2$, and the curve of it is symmetric about the **x**-axis.

And assuming $f(x) = \sin x + 2x + 1$, we get: $|y| = \sin x + 2x + 1$, and the curve of it is symmetric about the **x**-axis.

Suggestions or Solutions
To the **Problem** in the Example **2**

Put in a graph the equation as follows: $|y| = \sin |2x|$.

To begin with, putting in the *x-y* plane, the curve of the equation $y = \sin 2x$, we can put it the way below:

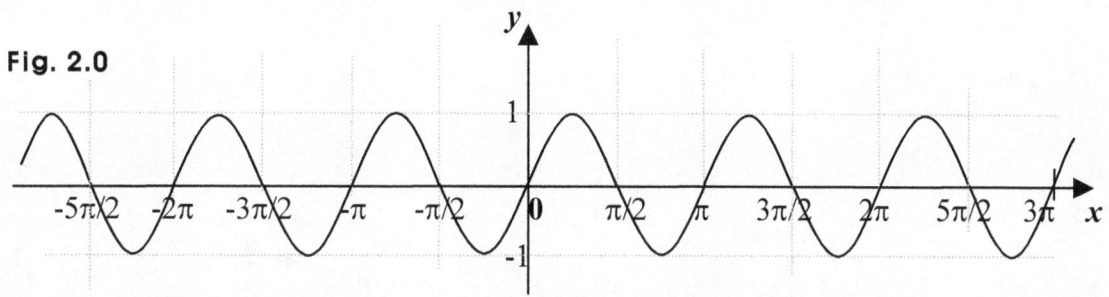

Fig. 2.0

Next, we know: $|y| \geq 0$ for all *y*, and $|2x| \geq 0$ for all *x*.

So we want to consider four different cases as follows:

First, assuming $x \geq 0$, and $y \geq 0$, we get: $y = \sin 2x$.

Next, assuming $x \geq 0$, and $y < 0$, we get: $-y = \sin 2x$, so we get: $y = -\sin 2x$.

Next, assuming $x < 0$, and $y < 0$, we get: $-y = \sin (-2x)$, so we get: $y = \sin 2x$.

And next, assuming $x < 0$, and $y \geq 0$, we get: $y = \sin (-2x)$, so we get: $y = -\sin 2x$.

Thus in sum, we have:

$y = \sin 2x$ for $x \geq 0$, and $y \geq 0$. $y = \sin 2x$ for $x < 0$, and $y < 0$.

$y = -\sin 2x$ for $x \geq 0$, and $y < 0$. $y = -\sin 2x$ for $x < 0$, and $y \geq 0$.

So beginning with the curve of $y = \sin 2x$ for $x \geq 0$, and $y \geq 0$, we get:

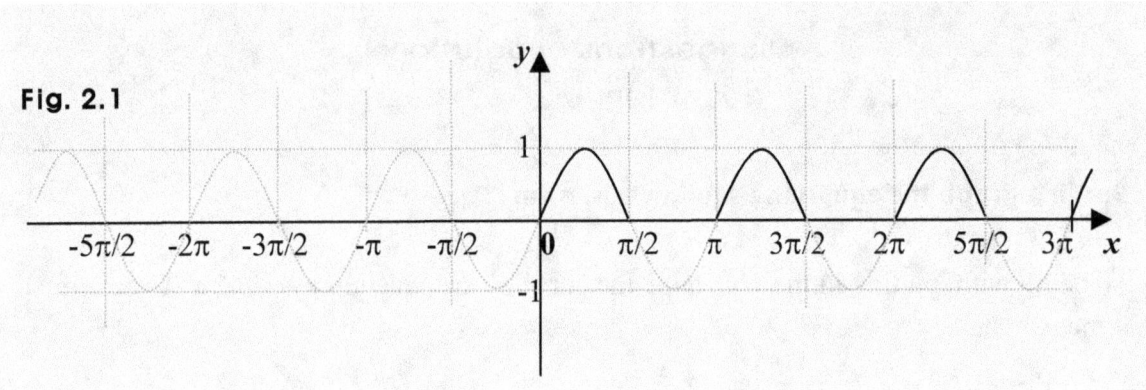

Fig. 2.1

Next, moving on to the curve of $y = \sin 2x$ for $x < 0$, and $y < 0$, we get:

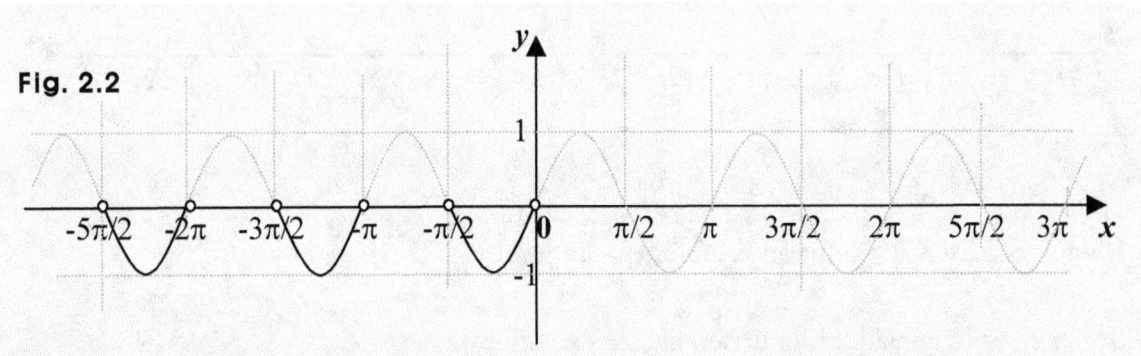

Fig. 2.2

Note that all the points $(n\pi/2, 0)$ for n integer ≤ 0, do not belong to the curve above.

Next, we can put in a graph, the curve of $y = -\sin 2x$ the way below:

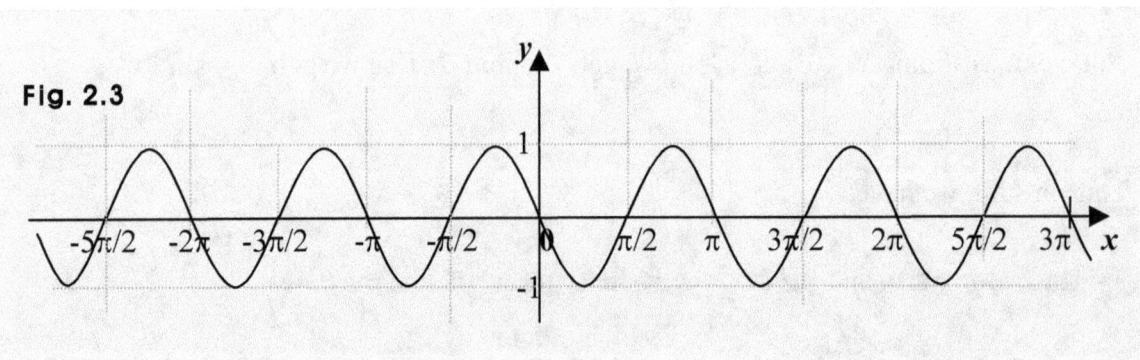

Fig. 2.3

So next, moving on to the curve of $y = -\sin 2x$ for $x \geq 0$, and $y < 0$, we get:

Fig. 2.4

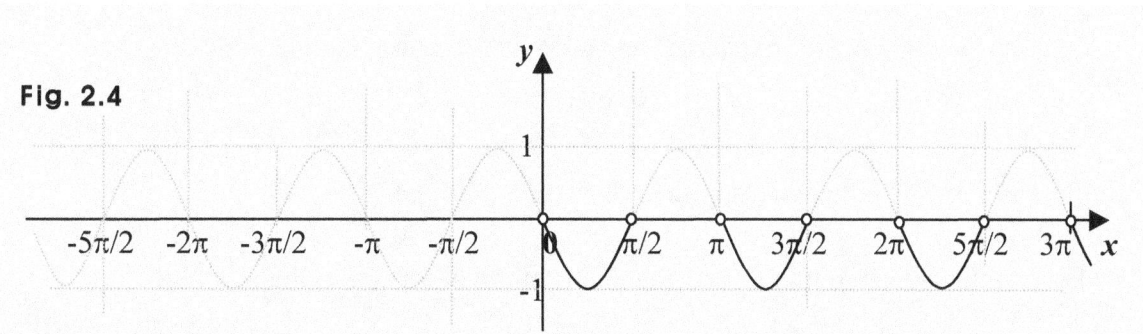

Note that all the points (**nπ/2, 0**) for **n** integer ≥ **0**, do not belong to the curve above.

And next, moving on to the curve of **y = -sin 2x** for **x < 0**, and **y ≥ 0**, we get:

Fig. 2.5

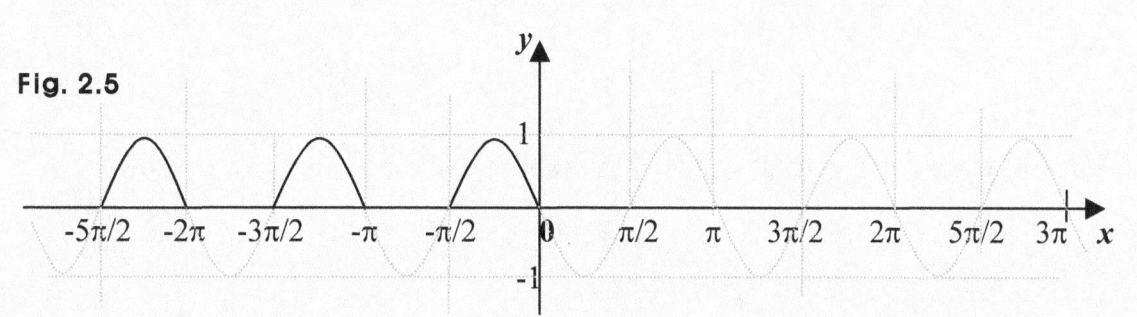

And thus, putting in a graph, the curve of |**y**| = **sin |2x|**, we can put it the way below:

Fig. 2.6

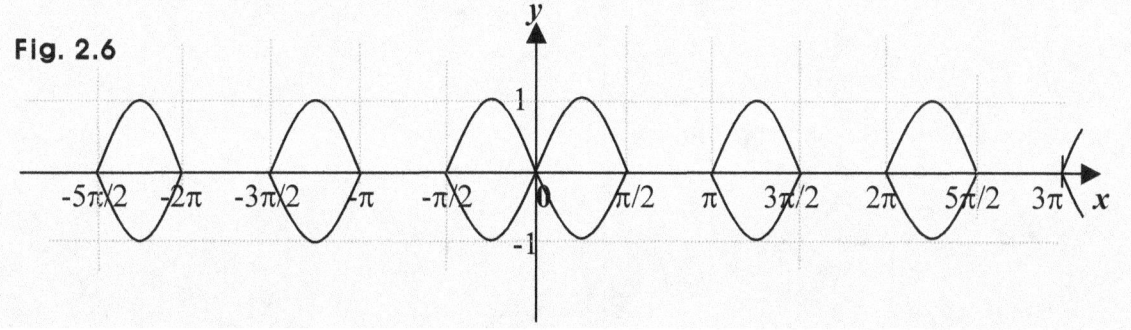

Notice that the curve is symmetric about the origin as well as the **x**-axis and the **y**-axis.

In fact, the curve of an equation |**y**| = **f**(|**x**|) is symmetric about the **x**-axis.

For instance, assuming $f(|x|) = |x|^2 - 3|x| + 2$, we get: $|y| = |x|^2 - 3|x| + 2$, and the curve of it is symmetric about the origin as well as the **x**-axis and the **y**-axis.

And assuming $f(|x|) = \sin|x| + 2|x| + 1$, we get: $|y| = \sin|x| + 2|x| + 1$, and the curve of it is symmetric about the origin as well as the **x**-axis and the **y**-axis.

Suggestions or Solutions
To the **Problem** in the Example **3**

Put in a graph the equation as follows: $y = \sin x + \sin |x|$.

To begin with, we can put in the *x-y* plane, the curve of $y = \sin x$, the way below:

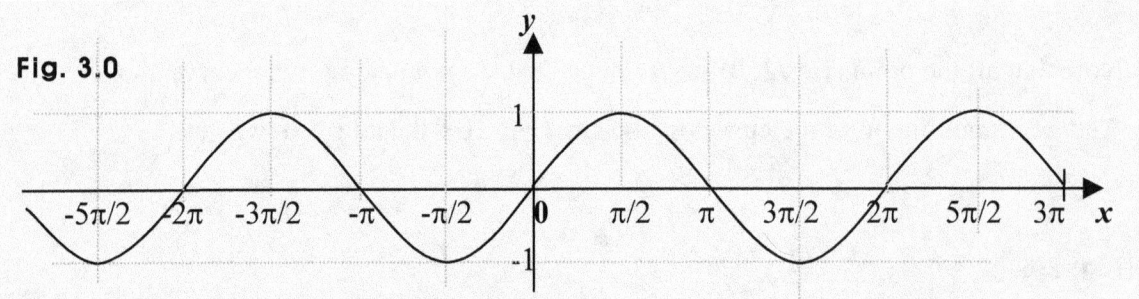

Fig. 3.0

And next, putting in a graph, the curve of $y = \sin |x|$, we can put it the way below:

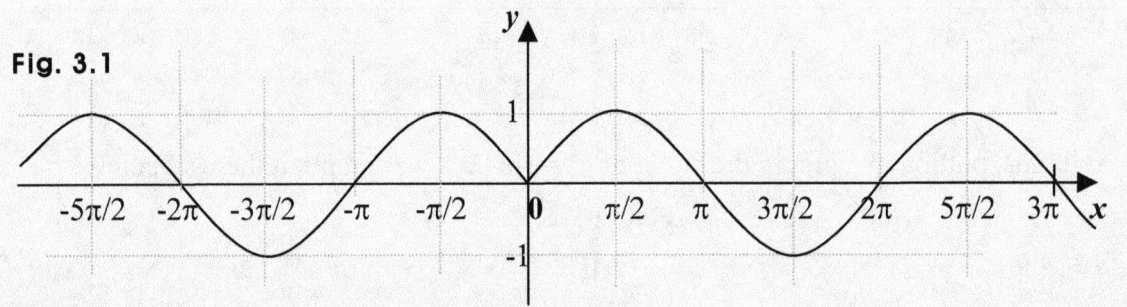

Fig. 3.1

So adding together the two curve above, we get:

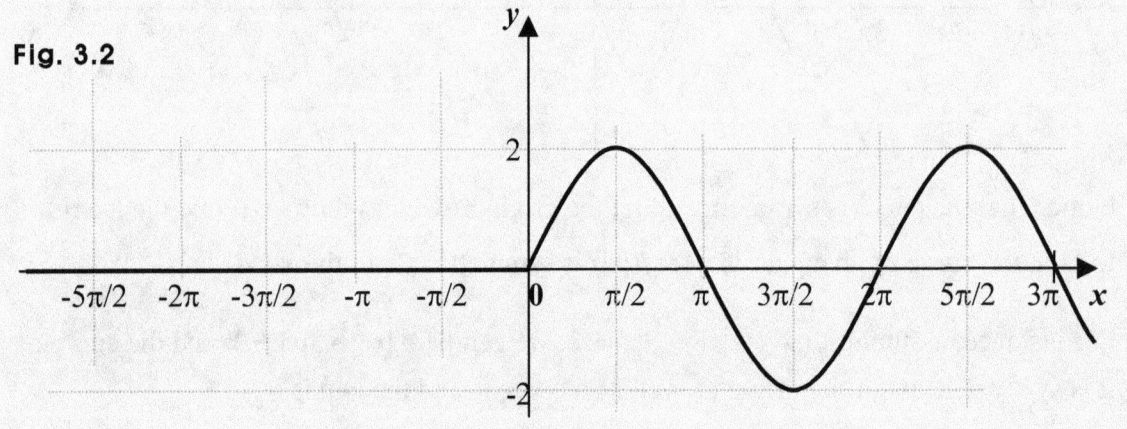

Fig. 3.2

And thus, the curve above is the curve of the equation $y = \sin x + \sin |x|$.

Let's take a look at now, how we can get the curve algebraically.

To begin with, we know: $|x| \geq 0$ for all x.

So we want to put the equation in two cases.

One is a case where $x \geq 0$, and the other is a case where $x < 0$.

And thus, removing the absolute sign, we get:

If $x \geq 0$, $y = \sin x + \sin |x| = \sin x + \sin x = 2\sin x$.

And if $x < 0$, $y = \sin x + \sin |x| = \sin x + \sin (-x) = \sin x - \sin x = 0$.

So we get: $y = 2\sin x$ for $x \geq 0$, and $y = 0$ for $x < 0$.

And thus, we can put in a graph, the curve of $y = \sin x + \sin |x|$ the way below:

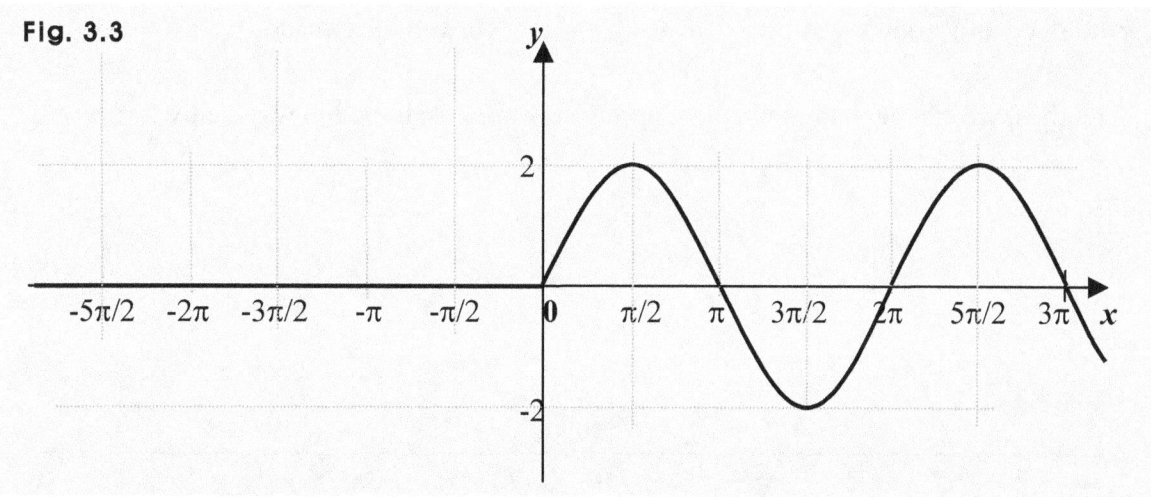

Fig. 3.3

And we can put the idea above the way below, too:

Assuming $f(x) = \sin x$, and $g(x) = \sin |x|$, we can set: $h(x) = f(x) + g(x)$.

And the curve of the function h is the same as that of the equation $y = \sin x + \sin |x|$.

So for each value of x, adding together the values of $f(x)$ and $g(x)$, we get the value of $h(x)$, which is in turn, the value of y in the equation $y = \sin x + \sin |x|$.

14

For instance, we get: $h(1) = f(1) + g(1)$ for $x = 1$.

And the y-value for $x = 1$ is: $\sin 1 + \sin |1| = \sin 1 + \sin 1 = 2\sin 1$.

And we get: $h(-1) = f(-1) + g(-1)$ for $x = -1$.

And the y-value for $x = -1$ is: $\sin (-1) + \sin |-1| = 0$, because $\sin (-1) = -\sin 1$.

So we can see that we get: $y = 2\sin x$ for $x \geq 0$, and $y = 0$ for $x < 0$.

What then, about the curve of $y = \sin x - \sin |x|$?

If $x \geq 0$, $y = \sin x - \sin |x| = \sin x - \sin x = 0$.

And if $x < 0$, $y = \sin x - \sin |x| = \sin x - \sin (-x) = \sin x + \sin x = 2\sin x$.

And thus, we can put in a graph, the curve of $y = \sin x - \sin |x|$ the way below:

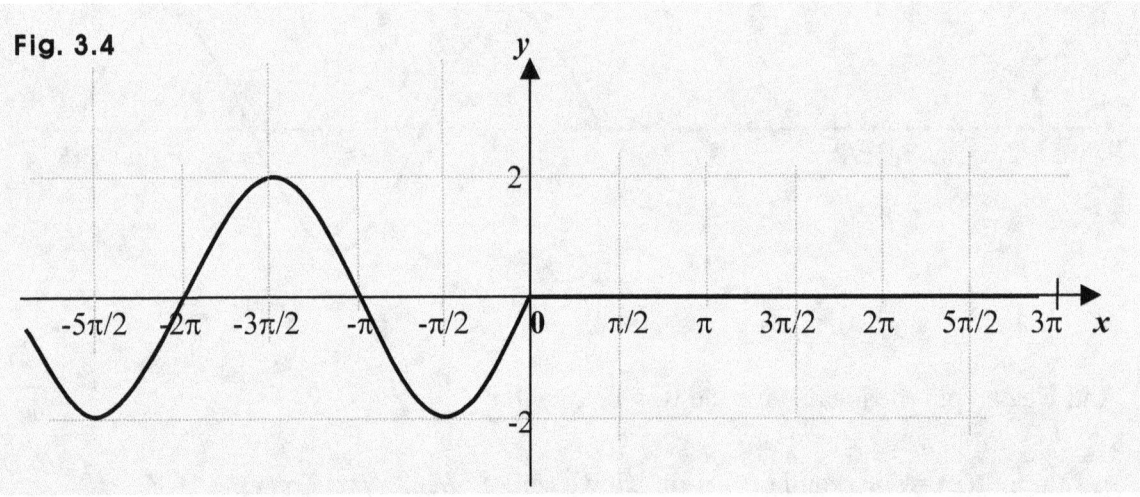

Fig. 3.4

What then, about the curve of $y = \sin^2 x$?

First, the curve of $y = \sin x$ is as follows:

Fig. 3.5

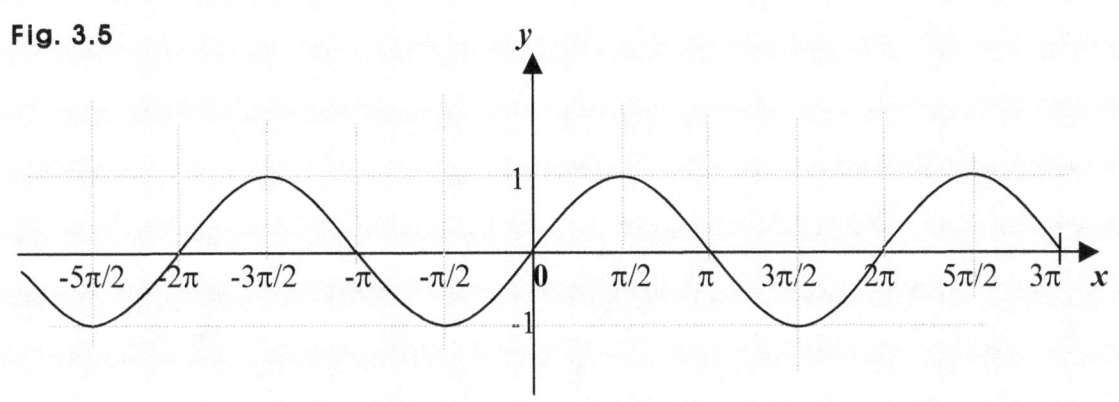

So we can expect the curve of $y = \sin^2 x$ to be made the way below:

Fig. 3.6

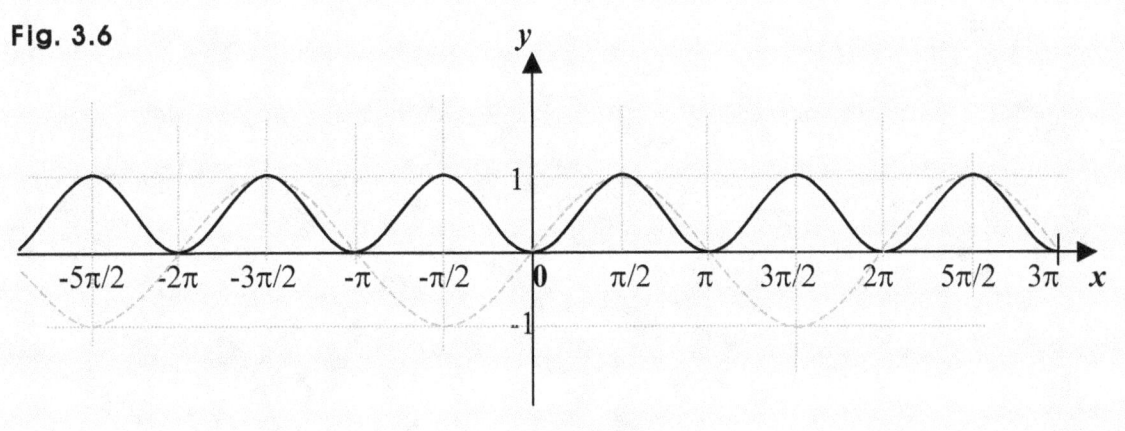

B. Cosine Functions

What is a cosine function?

It is a function of which each output is a trig-ratio called the cosine. So for instance, putting a cosine function in the *x-y* coordinate system, we can put it the way below:

$y = f(x) = \cos x$, where *x* is an angle.

So like a sine function, it is a function of an angle, and thus, takes an angle as an input.

And such an angle is a governing angle. So the input variable *x* takes a governing angle, which is an angle made by the terminal ray turning about the origin in the *x-y* plane.

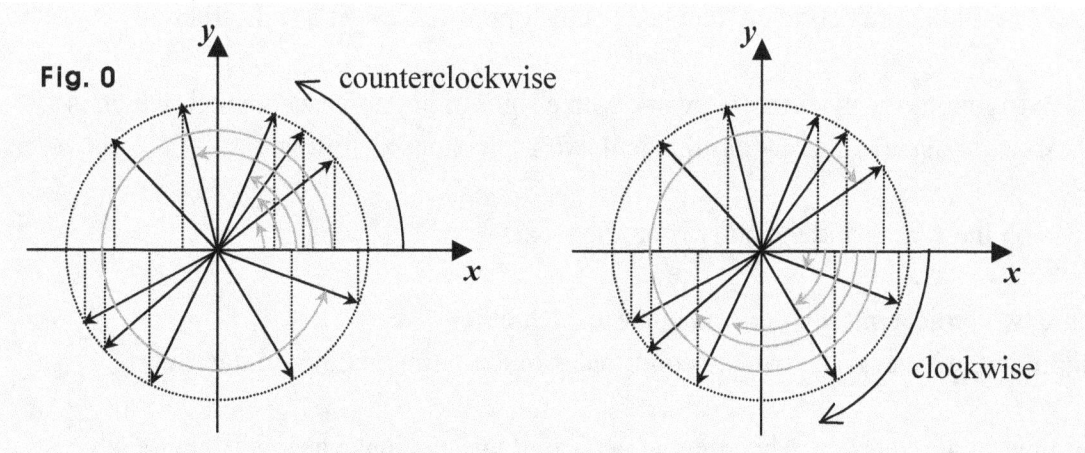

And assuming <u>the length of the ray is 1</u>, the terminal point is (x, y), we can see a lot of right triangles. In each, the ray is the hypotenuse, *x* is the adjacent, and *y* is the opposite.

So again, in the *x-y* plane where the ray is turning counterclockwise, placing four lamps the way below, we get shadows of the ray (the projections) on both axes.

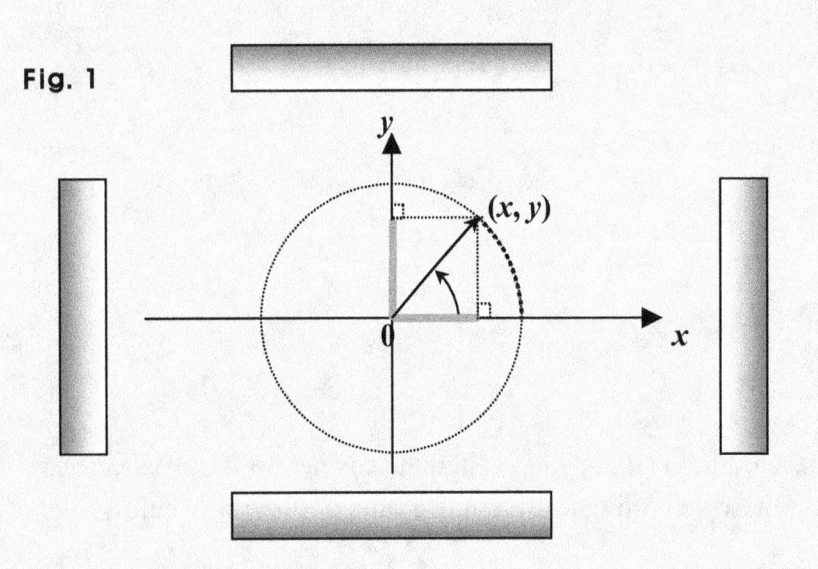

Fig. 1

Then, as the ray keeps turning, we can see on one axis, a linear motion where the shadow (projection) decreases, while on the other axis, another linear motion where the shadow increases. And such linear motions keep repeating as the ray keeps turning.

So at every moment, assuming the ray is a hypotenuse, the projection on the *x*-axis is the adjacent, and the projection on the *y*-axis is the opposite, we get a right triangle.

And in trigonometry dynamic, we work with a right triangle transcendental, which is a right triangle made of the ray turning, and two projections as follows:

One is on the *x*-axis, and the other is on the *y*-axis.

So the two projections are perpendicular to each other.
And in fact, we use as the two the coordinates of the terminal point in the ray.

And thus, at any moment while the ray is turning, we can make a right triangle, where the hypotenuse is the ray, the adjacent is the projection on the *x*-axis, that is, the *x*-coordinate at the terminal point, and the opposite is the projection on the *y*-axis, that is, the *y*-coordinate at the terminal point.

And depending on the amount of angle made by the ray, the projections get determined. In short, the angle made determines the projections. So what makes such a right triangle is in fact, the ray turning and the amount of angle made by the ray.

In fact, since the length of the ray is constant, the length of the hypotenuse is constant, too, so the angle made by the ray turning determines a right triangle, because that angle determines the opposite and the adjacent. And we use that angle to get the trig-ratios. And thus, each angle made by the ray turning is a governing angle.

And when taking trig-ratios in a right triangle transcendental, too, we take the ratios the way we take them in a right triangle normal.
So it is always the case where the cosine is: the adjacent over the hypotenuse, the sine is the opposite over the hypotenuse, and the tangent is: the opposite over the adjacent.

• Now, in dynamic trigonometry, we often work with functions called trig-functions. Of each, each input is a governing angle, which is made by the ray turning in the *x-y* plane.

So in the cosine function, $y = f(x) = \cos x$, too, the input variable *x* gets a governing angle. Usually though, we just call governing angles, angles, and use numbers as angles, since those angles are in radian.

So for instance, assuming *f* is a cosine function, and is: **cos *x***, and the domain is a set of all angles, we say that the domain is a set of all real numbers, and can put the trig-function *f* this way: $y = f(x) = \cos x$ for *x* real, or simply this way, too: $y = f(x) = \cos x$.

What then, about outputs?

They are trig-ratios, because in the cosine function $y = f(x) = \cos x$, the *x* gets an angle, so each value of **cos *x***, that is, each value of *f(x)* is a trig-ratio, which is a number.
What then, about the range?

The range is a set of all numbers from -1 to 1, and each of the numbers is a trig-ratio, which is an output, which is the value of *f(x)*, which is the value of *y*.
So the range of the cosine function *f* can be put this way, too: $-1 \leq y \leq 1$, or $|y| \leq 1$.
And we will see now, how it is the case.

So let's first, get back to the *x-y* plane where the projection is made on the *x*-axis.

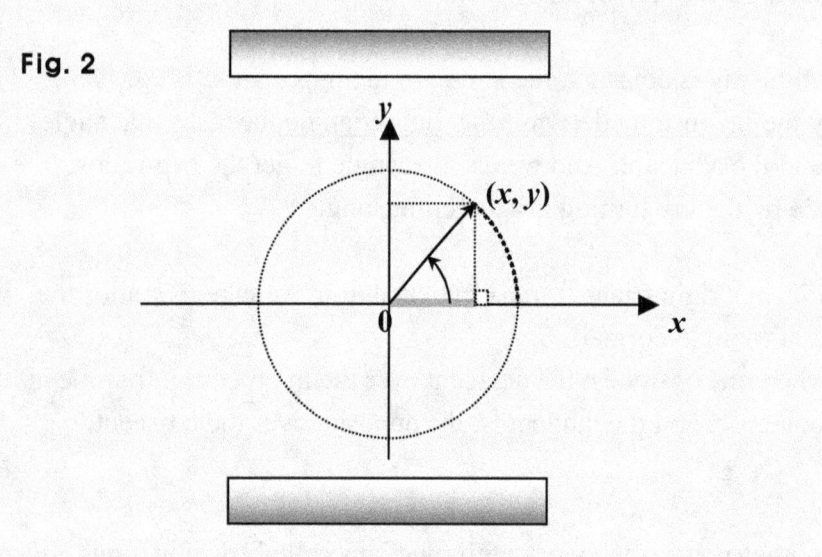

Fig. 2

Suppose that <u>the ray is of length 1</u>, and is placed on the *x*-axis to the right of the origin.

Suppose now that the ray starts turning counterclockwise.

Then, from above the *x*-axis, we can see that (the length of) the projection on the *x*-axis starts decreasing from 1.

Even before the ray turns, assuming the ray is a hypotenuse, we can get a right triangle. What right triangle?

It is a right triangle transcendental, where the hypotenuse is the ray, and the adjacent is the projection on the *x*-axis, and is the ray itself, since the ray is on the *x*-axis.

So when the ray is at rest, the projection on the *x*-axis is no other than the ray, so the adjacent is 1, too, which is impossible though, in a triangle normal. So we can call it a right triangle transcendental.

And we know that the angle made is 0 before the ray turns, and that by definition, <u>the cosine is: the adjacent over the hypotenuse.</u>   So we get: **cos 0 = 1/1 = 1**.

Suppose now that the angle made is θ, and that the terminal point in the ray is (x, y).

Then, we can get a right triangle transcendental, where the hypotenuse is the ray, the adjacent is x, and the opposite is y.   What then, is the cosine of the angle θ?

We know by definition, <u>the cosine is: the adjacent over the hypotenuse</u>. And we know the hypotenuse is 1, since the ray is 1, and the adjacent is x.   So we get: $\cos \theta = x/1 = x$.

•• So for any angle θ, assuming the ray is 1, that is, the hypotenuse is 1, and the terminal point is (x, y), we get: $\cos \theta = x$, which is of course, the x-coordinate at the point (x, y).

Suppose now that the angle θ is $\pi/2$.   What right triangle then, can we get?

It is a right triangle transcendental, where the hypotenuse is the ray, which is 1, and the adjacent is the projection on the x-axis, which is now 0.   So we get: $\cos \pi/2 = 0$.

And we know if θ is $\pi/2$, the ray is on the y-axis, above the origin, the ray is of length 1, and the terminal point is (x, y).   So what is the x-value at (x, y) if θ is $\pi/2$?

It is 0.   So the x-value at (x, y) is the value of the adjacent, which is 0 when θ is $\pi/2$.

Suppose next that the angle θ is π.   What right triangle then, can we get?

It is a right triangle transcendental, where the hypotenuse is the ray, which is 1, and the adjacent is the projection on the x-axis, which is now -1.   So we get: $\cos \pi = -1$.

And we know if θ is π, the ray is on the x-axis, to the left of the origin, the ray is of length 1, and the terminal point is (x, y).   So what is the x-value at (x, y) if θ is π?

It is -1. So the x-value at (x, y) is the value of the adjacent, which is -1 when θ is π.

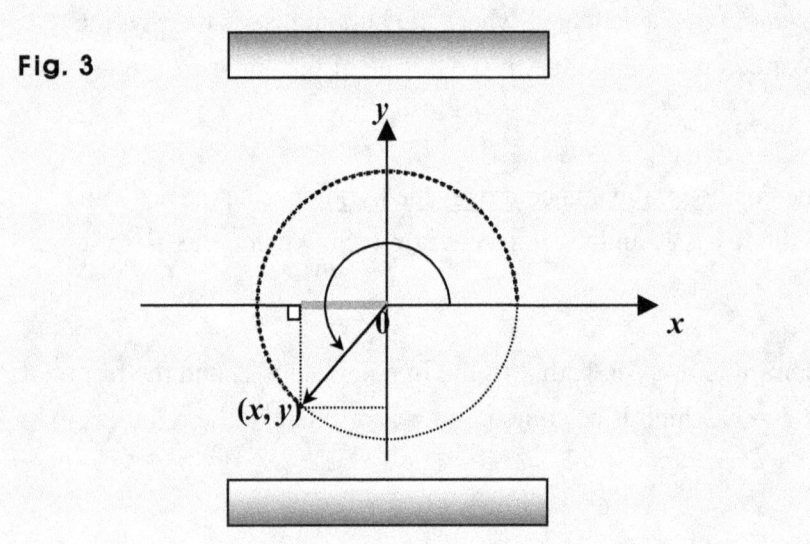

Fig. 3

Suppose now that the angle θ is $3\pi/2$. What right triangle then, can we get?

It is a right triangle transcendental, where the hypotenuse is the ray, which is 1, and the adjacent is the projection on the x-axis, which is now 0. So we get: **$\cos 3\pi/2 = 0$**.

And we know if θ is $3\pi/2$, the ray is on the y-axis, below the origin, the ray is of length 1, and the terminal point is (x, y). So what is the x-value at (x, y) if θ is $3\pi/2$?

It is 0. So the x-value at (x, y) is the value of the adjacent, which is 0 when θ is $3\pi/2$.

What then, are the maximum and the minimum of the x-value?

The maximum is 1, and the minimum is -1, which is because the terminal point makes a circle of radius 1 centered at the origin if the ray makes a complete turn.

So now, as the governing angle θ changes from 0 to 2π, what value does **cos θ** get?

If the ray is 1, the hypotenuse is 1, so the value of **cos θ** is the x-value at the terminal point (x, y), and x covers all the numbers from -1 to 1.

So if $0 \leq \theta \leq 2\pi$, we get: **$-1 \leq \cos \theta \leq 1$**, that is, **$|\cos \theta| \leq 1$**.   What if $2\pi \leq \theta \leq 4\pi$?

If $2\pi \leq \theta \leq 4\pi$, we get: **$-1 \leq \cos \theta \leq 1$**, too, that is, **$|\cos \theta| \leq 1$**.

And thus, if $0 \leq \theta \leq 2n\pi$, where n is an integer ≥ 1, we get: **$|\cos \theta| \leq 1$**.

That is, no matter how many complete turns the ray may make, we get: **$|\cos \theta| \leq 1$**.

Suppose now again, the ray of length 1 is at rest on the x-axis to the right of the origin.

• Suppose this time, the ray turns clockwise.

Then, from below the x-axis, we can see that at the beginning, the projection on the x-axis starts decreasing from 1.

Suppose now that the angle made is θ, and that the terminal point in the ray is (x, y).

What right triangle then, can we get?

It is a right triangle transcendental, where the hypotenuse is the ray, the adjacent is x, and the opposite is y.   What then, is the cosine of the angle θ?

We know <u>the cosine is: the adjacent over the hypotenuse</u>, the hypotenuse is 1, and the adjacent is x.   So we get: **$\cos \theta = x/1 = x$**.   Note that the angle θ is negative now.

24

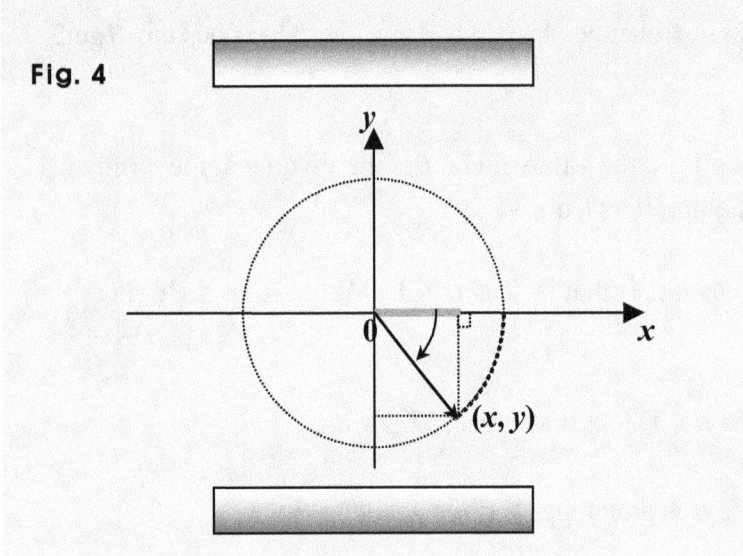

Fig. 4

Suppose now that the angle θ is -π/2.   What right triangle then, can we get?

It is a right triangle transcendental, where the hypotenuse is the ray, which is 1, and the adjacent is the projection on the *x*-axis, which is now 0.   So we get: **cos (-π/2) = 0**.

And we know if θ is -π/2, the ray is on the *y*-axis, below the origin, of course, the ray is of length 1, and the terminal point is **(x, y)**.   So what is the *x*-value at **(x, y)** if θ is -π/2?

It is 0.   So the *x*-value is the value of the adjacent, which is 0 when θ is -π/2.

Suppose now that the angle θ is -π.   What right triangle then, can we get?

It is a right triangle transcendental, where the hypotenuse is the ray, which is 1, and the adjacent is the projection on the *x*-axis, which is now -1.   So we get: **cos (-π) = -1**.

And we know if θ is -π, the ray is on the *x*-axis, to the left of the origin, the ray is of length 1, and the terminal point is **(x, y)**.   So what is the *x*-value at **(x, y)** if θ is -π?

It is -1. So the x-value is the value of the adjacent, which is -1 when θ is -π.

Suppose now that the angle θ is -3π/2.   What right triangle then, can we get?

It is a right triangle transcendental, where the hypotenuse is the ray, which is 1, and the adjacent is the projection on the x-axis, which is now 0.   So we get: **cos (-3π/2) = 0**.

And we know if θ is -3π/2, the ray in on the y-axis, above the origin, of course, the ray is of length 1, and the terminal point is (x, y). So what is the x-value at (x, y) if θ is -3π/2?

It is 0. So the x-value is the value of the adjacent, which is 0 when θ is -3π/2.

And thus, the maximum of the x-value is 1, and the minimum is -1, which is also because the terminal point makes a circle of radius 1 centered at the origin if the ray makes a complete turn.

So now, as the governing angle θ changes from 0 to -2π, what value does **cos θ** get?

If the ray is 1, the hypotenuse is 1, so the value of **cos θ** is the x-value at the terminal point (x, y), and x covers all the numbers from –1 to 1.

So if **-2π ≤ θ ≤ 0**, we get: **-1 ≤ cos θ ≤ 1**, that is, **|cos θ| ≤ 1**.   What if **-4π ≤ θ ≤ -2π**?

If **-4π ≤ θ ≤ -2π**, we get: **-1 ≤ cos θ ≤ 1**, too, that is, **|cos θ| ≤ 1**.

And thus, if **2nπ ≤ θ ≤ 0**, where **n** is an integer ≤ **-1**, we get: **|cos θ| ≤ 1**.

So now, putting threads together, if the ray is 1, and (x, y) is the terminal point, we get: **cos θ = x** for any angle θ, and get: **|cos θ| ≤ 1**, since $|x| ≤ 1$.

It's because <u>the cosine is: the adjacent over the hypotenuse</u>, the hypotenuse is 1, since the ray is 1, and the adjacent is the *x*-value at the terminal point (*x, y*), which makes a circle of radius 1 centered at the origin if the ray makes a complete turn.

••• So using the facts above, we can form a function of the angle θ.   How?

As the angle θ changes, the cosine trig-ratio **cos** θ changes. So we can take as an input each angle that θ gets, and take as an output each trig-ratio that **cos** θ produces.

Assuming thus, the function is *f*, the domain is a set of all angles, and *c* is the output variable, we can set: $c = f(\theta) = \textbf{cos } \theta$.   And we call the function *f* a cosine function.

• What then, about the curve of the cosine function $c = f(\theta) = \textbf{cos } \theta$?

Assuming the ray is 1, and the terminal point is (*x, y*), we get: **cos** $\theta = x$. So the *x*-value at the terminal point (*x, y*) in the ray that makes the angle θ is the value of **cos** θ.

In short, the *x*-value at the terminal point is: **cos** θ.

So this time, rotating by π/2 counterclockwise about the origin, the *x-y* plane where the ray is turning, we can get the curve of the cosine function $f(\theta)$ the way as follows:

Fig. 5

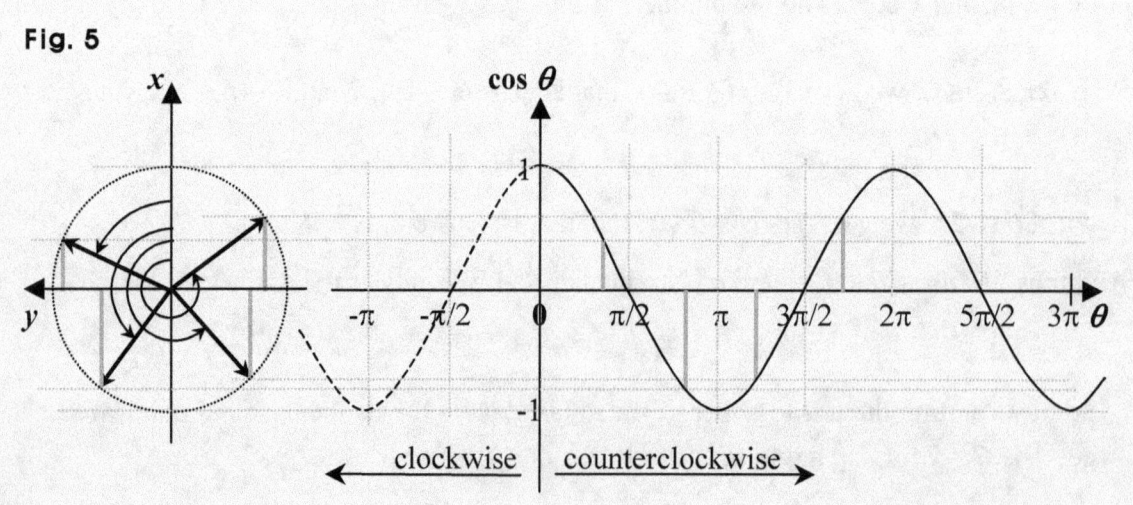

And of course, we can put in the *x-y* system, the function $c = f(\theta) = \cos\theta$.
Simply replacing *c* with *y*, and θ with *x*, we get: $y = f(x) = \cos x$.

And we can use numbers as angles. So if the domain is a set of all angles, we usually say that the domain is a set of all real numbers, and we can just set: $y = f(x) = \cos x$ for *x* real.

And just setting: $y = f(x) = \cos x$, we mean that the domain is a set all real numbers.

And we can put in the *x-y* plane, the curve of the cosine function above the way below.

Fig. 6 $y = f(x) = \cos x$

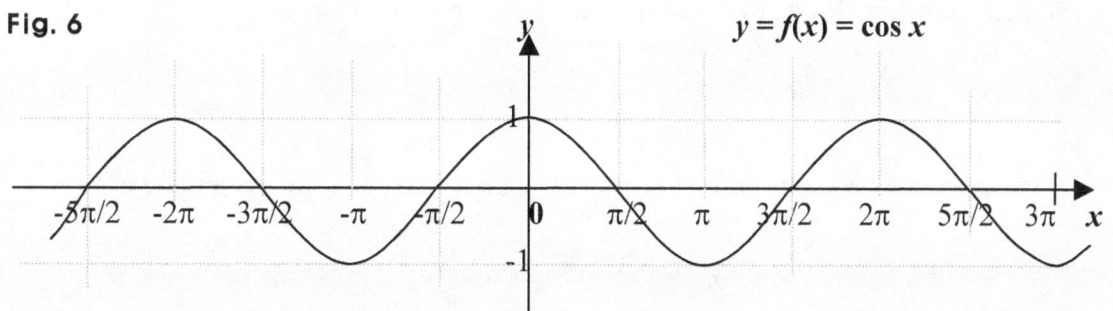

The curve above is called the cosine curve, and is the same as the curve in **Fig. 5** above.

And we know that the output can be every value from -1 to 1.
So the range is: $-1 \leq y \leq 1$, that is, $|y| \leq 1$.

And of course, if the domain is not a set of all real numbers, the range can be other than $|y| \leq 1$.   If for instance, $y = g(x) = \cos x$ for $-\pi/2 \leq x \leq \pi/2$, the range is: $0 \leq y \leq 1$.

And the cosine function $y = g(x) = \cos x$ can be called the prototype, and thus, is in the most basic form. Assuming *F* is a cosine function, too, where the domain is a set of all real numbers (that is, all angles), and using a general form, we can put it the way below:

$y = F(x) = A \cdot \cos \{w(x + a)\} + b$ for *x* real, where *A*, *w*, *a*, and *b* are constant.

And we can just put it this way, too: $y = F(x) = A \cdot \cos w(x + a) + b$ for *x* real.

Or more simply this way, too: $y = F(x) = A \cdot \cos w(x + a) + b$.

(Note that $w(x + a)$ represents an angle, but *A* and *b* represent just a number each.)

Then, the range is a set of all numbers from $-|A| + b$ to $|A| + b$.

So the range of F can be put this way, too: $-|A| + b \leq y \leq |A| + b$.

That is, the curve of F is bounded by the interval where $-|A| + b \leq y \leq |A| + b$.

Then, $2|A|$ is the width of the curve (or wave), and $|A|$ is called the amplitude.

So the amplitude indicates half the width of the curve (or wave).

What then, is the amplitude of the cosine function $y = f(x) = \cos x$?

Since the curve of f is bounded by the interval where $-1 \leq y \leq 1$, the amplitude is 1.

And $|w|$ is called the frequency, $\frac{2\pi}{|w|}$ is the period, and a is called the phase.

So in the function $y = f(x) = \cos x$, the phase is 0, and the frequency is 1, so the period is 2π (i.e., $360°$).   How come though?

The prototype in cosine functions can be put this way: $y = f(x) = \cos x$ for x real.
And using a general form, we can put a cosine function called F the way below:

$y = F(x) = A \cdot \cos w(x + a) + b.$

Then, a is called the phase, $|w|$ is called the frequency, and $\frac{2\pi}{|w|}$ is the period. And we can put the prototype into the general form this way: $y = f(x) = 1 \cdot \cos 1(x + 0) + 0.$

So the phase is 0, and the frequency is 1, so the period is 2π.
What do we mean by the period though?

Putting in the x-y plane, the curve of the cosine function f above, we get:

Fig. 7

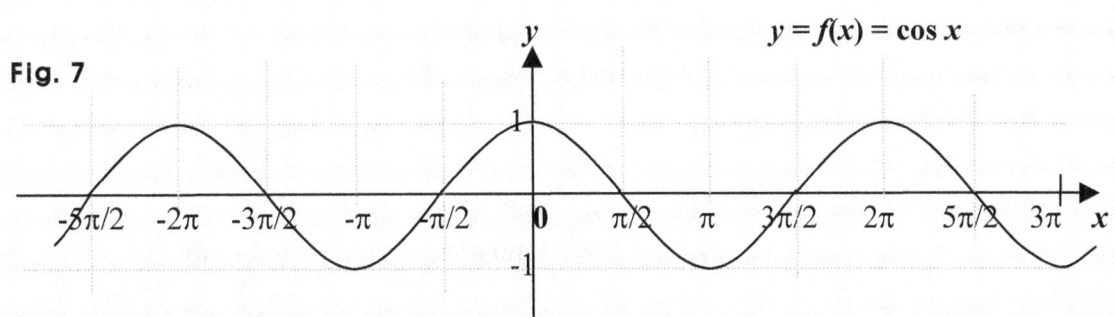

$y = f(x) = \cos x$

Then, we can see that a part of the cuve repleats itself.
And the part is in fact, the smallest part repeating itself.   What then, is the part?

It is the part from 0 to 2π. So the part repeats itself every 2π interval. And we call such an interval a period.   So the interval 2π is the period in the function $y = f(x) = \cos x$.

• And thus, we call a cosine function a *periodic* function. And of course, the same is true, too, for sine functions, tangent functions, and other trig-functions as cosecant functions.

And the cosine function *f* above is the prototype, and thus, is in the most basic form.

  • So the period **2π** can be called the *basic period* in *cosine functions*. In other words:
  • The interval **2π** can be called the *basic interval* in *cosine functions*.

And for instance, in the general form, setting *w* to 2, *a* and *b* to 0 each, and *A* to 1, we get a new function where: $y = g(x) = \cos 2x$ for *x* real.

And putting in the *x-y* plane, the curve of the cosine function *g* above, we get:

Fig. 8

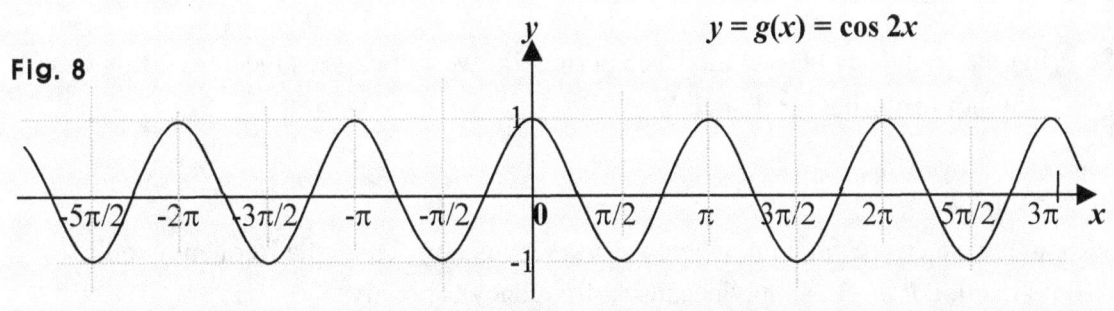

$y = g(x) = \cos 2x$

Then again, we can see that a part of the cuve repleats itself.
And the part is of course, the smallest part repeating itself.   What then, is the part?

It is the part from 0 to π.   So the part repeats itself every π interval.
And we call such an interval a period.   So the interval π is the period in the function g.

How many times then, does the smallest part stated above appear in the basic interval (the basic period) 2π?

It appears twice in the basic interval 2π, which is the basic period. Then, we say that the frequency is 2 in the function g. What then, do we mean by the frequency?

It is the number of times the the period appears in the basic period 2π, which is the basic interval.   In short, the frequency is the number of times the period appears in 2π.

And the same is true for sine functions, too.
The basic period however, in tangent functions is not 2π but π.   So in general, the frequency in a trig-function is the number of times its period appears in its basic period.

And thus, in the general form $A \cdot \cos w(x + a) + b$, $|w|$ is the frequency. Then, given a cosine function in the general form, how can we get its period?

We know that $|w|$ is the frequency, that is, the number of times the period appears in the basic period 2π.

So using the frequency $|w|$ and the basic period 2π, we can express the period this way: $\frac{2\pi}{|w|}$. Why not just w but $|w|$ though?

We can have a cosine function where $y = h(x) = \cos(-x)$. Then, the frequency in the cosine function h is $|-1|$, simply because a frequency is positive.

We can put **h** this way, too, though: $y = h(x) = \cos x$, because: $\cos(-x) = \cos x$.   Why?

Assuming the ray, that is, the hypotenuse is of length 1, we get:

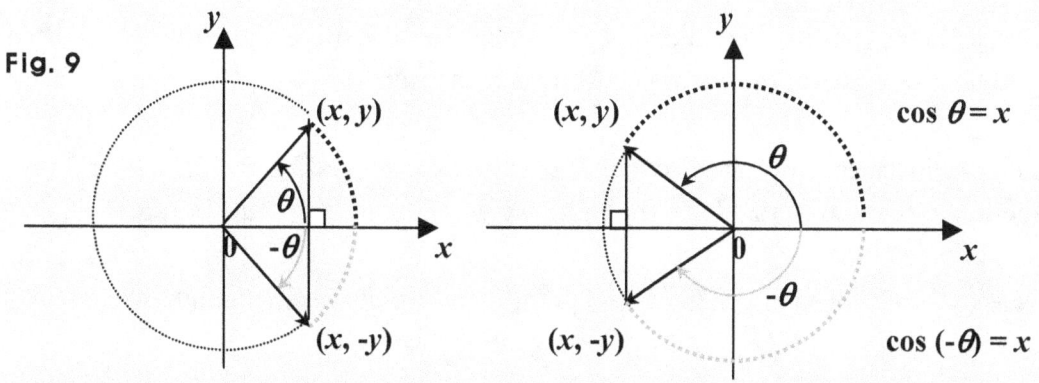

Fig. 9

So we get: $\cos(-\theta) = \cos\theta$.   And putting in the *x-y* plane, the curve of **h**, we get:

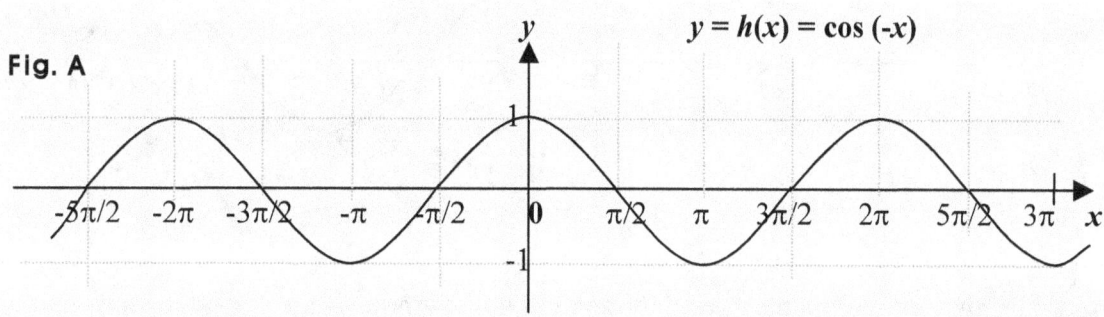

Fig. A $y = h(x) = \cos(-x)$

So we get the same curve as the one for $y = f(x) = \cos x$, and the period of **h** is 2π, too.

• And next, what do we mean by the phase?

Using the general form, we can put a cosine function called **F** the way below:

$y = F(x) = A \cdot \cos w(x + a) + b$ for **x** real, where **A**, **w**, **a**, and **b** are constant.

Then, **a** is called the phase, |**w**| is called the frequency, and $\frac{2\pi}{|w|}$ is the period.

And next, assuming: $y = Q(x) = A \cdot \cos wx + b$ for x real, and shifting the curve of the function $Q(x)$ by $-a$ in the direction of the x-axis, we get the curve of the function F.

So the two curves themselves of F and Q are the same, and moving the curve of Q in the amount of $-a$ along the x-axis, we get the curve of F.

And thus, as in the case of sine functions, if the phase a is positive, the curve gets shifted to the left, and if negative, the curve gets shifted to the right.

Assuming for instance, shifting the curve of $p(x) = \cos x$ for $-3\pi/2 \leq x \leq 5\pi/2$ by $\pi/2$ in the direction of the x-axis, that is, to the right, we get the curve of a function below:

$q(x) = \cos (x - \pi/2)$ for $-\pi \leq x \leq 3\pi$.   And putting the two curves in a graph, we get:

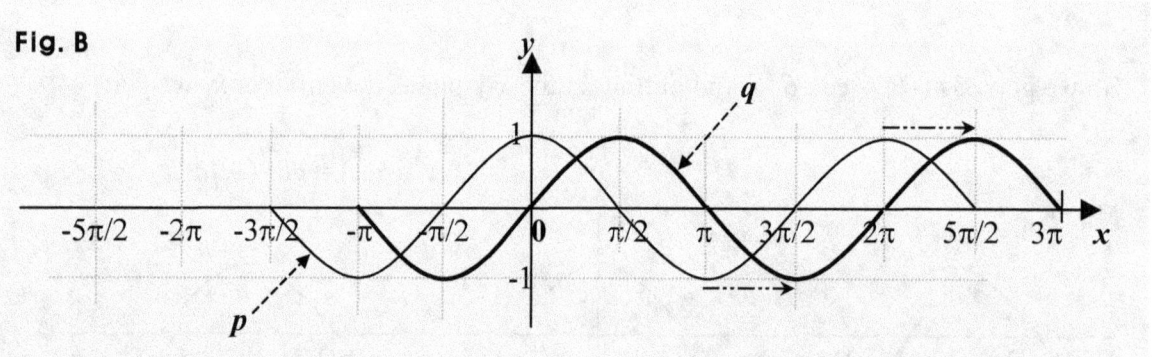

Fig. B

And thus, the curve of p gets moved to the right in the amount of $\pi/2$. And the new curve is the curve of q. So we can call the phase a horizontal shift, too.

And notice that the curve of q is a part of the curve of the sine function $s(x) = \sin x$.

What then, about the constant b in $y = Q(x) = A \cdot \cos wx + b$?

It is called a vertical shift. So shifting the curve of $y = V(x) = A \cdot \cos w(x + a)$ by b in the direction of the y-axis, we get the curve of the function F above.

So the two curves themselves of *F* and *V* are the same, and moving the curve of *V* in the amount of *b* along the *y*-axis, we get the curve of *F*.

Assuming thus, for instance, $u(x) = \cos x$ for *x* real, and shifting the curve of $u(x)$ by 1 in the direction of the *y*-axis, we get the curve of a function: $v(x) = \cos x + 1$ for *x* real.

So putting the two functions *u* and *v* in a graph, we can get:

Fig. C

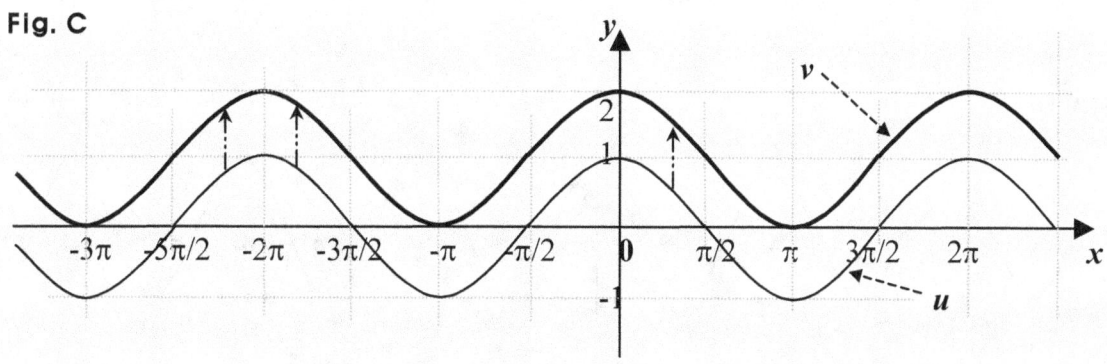

So the curve of *u* gets moved upward in the amount of 1. And the new curve is the curve of *v*. Thus, we can call the *b* the amount of a vertical shift. In short, *b* is a vertical shift.

And let's now put in a graph, for instance, the curve of a cosine function below:

$y = C(x) = -2\cos(-2x + \pi) + 1$ for *x* real.

To begin with, we have: $\cos(-\theta) = \cos\theta$.

So we can get: $\cos(-2x + \pi) = \cos\{-(2x - \pi)\} = \cos(2x - \pi)$.

Thus, we get: $-2\cos(-2x + \pi) + 1 = -2\cos(2x - \pi) + 1$.

And putting it in the general form, we get: $-2\cos 2(x - \pi/2) + 1$.

So we can now see that the amplitude is 2, the frequency is 2, the phase is $-\pi/2$, that is, shifting by $\pi/2$ to the right, and the vertical shift is 1.

And next, putting it in a graph, we may want to begin with the prototype: **cos *x***.

First, setting the frequency to 2, we get: **cos 2*x***. And we know the period is: the basic period over the frequency, and that the basic period for the cosine is 2π.

So we can see that the period is: $2\pi/2 = \pi$.

And thus, we can put the curve with **cos 2*x*** the way below:

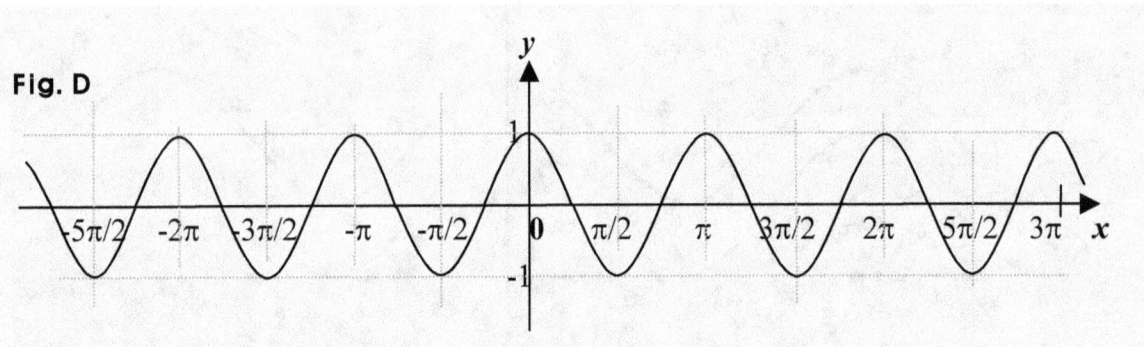

Fig. D

Next, setting the amplitude to 2, we get: **2 cos 2*x***.

We know that the amplitude indicates half the width of the curve.

And thus, we can put in a graph, the curve with **2 cos 2*x*** the way below:

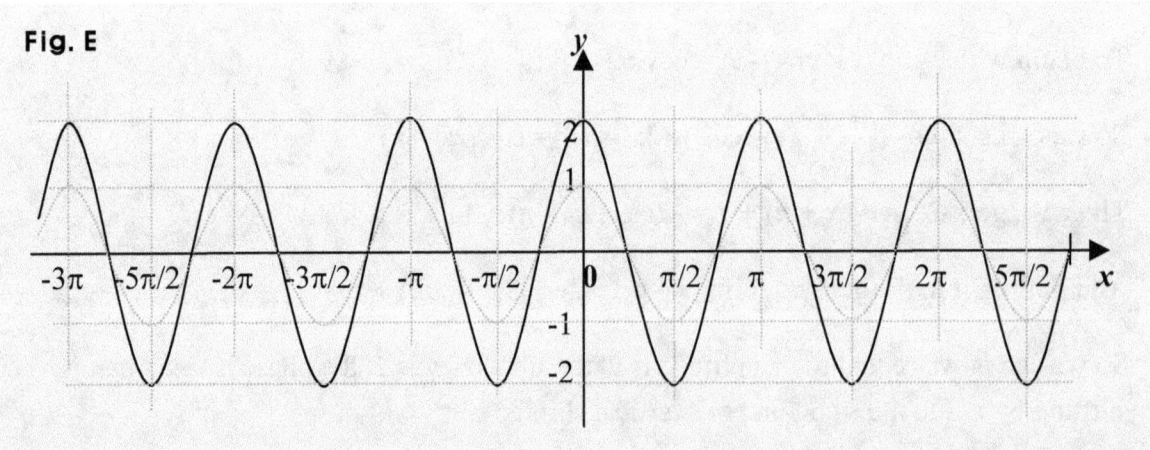

Fig. E

However, we want to put the negative sign in front. So we get: **-2 cos 2x**.

Fig. F

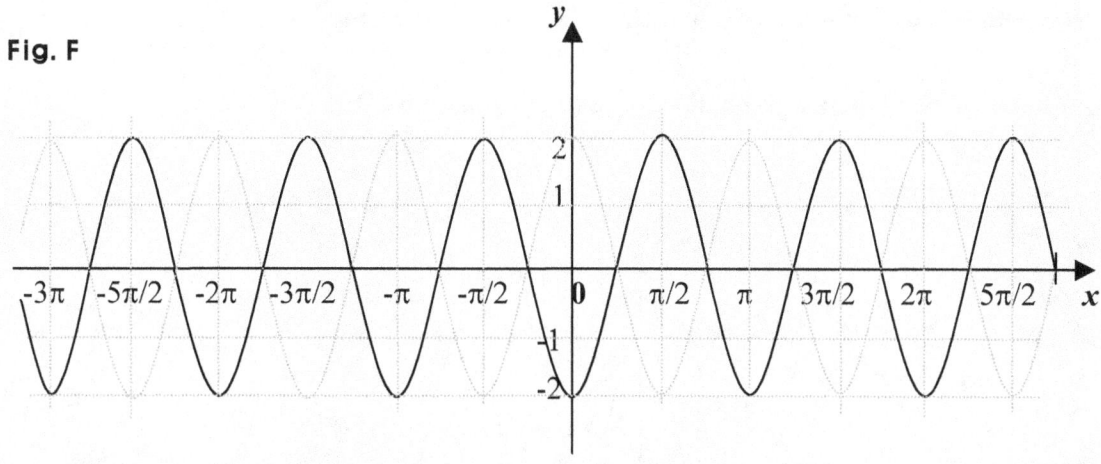

Next, setting the phase to –π/2, we get: **-2 cos 2(x – π/2)**.

Shifting the curve with **–2 cos 2x** by π/2 in the direction of the *x*-axis, we get the curve with **–2 cos 2(x – π/2)**.

So moving the curve with **–2 cos 2x** in the amount of π/2 to the right, we get the curve with **–2 cos 2(x – π/2)**, which is below:

Fig. G

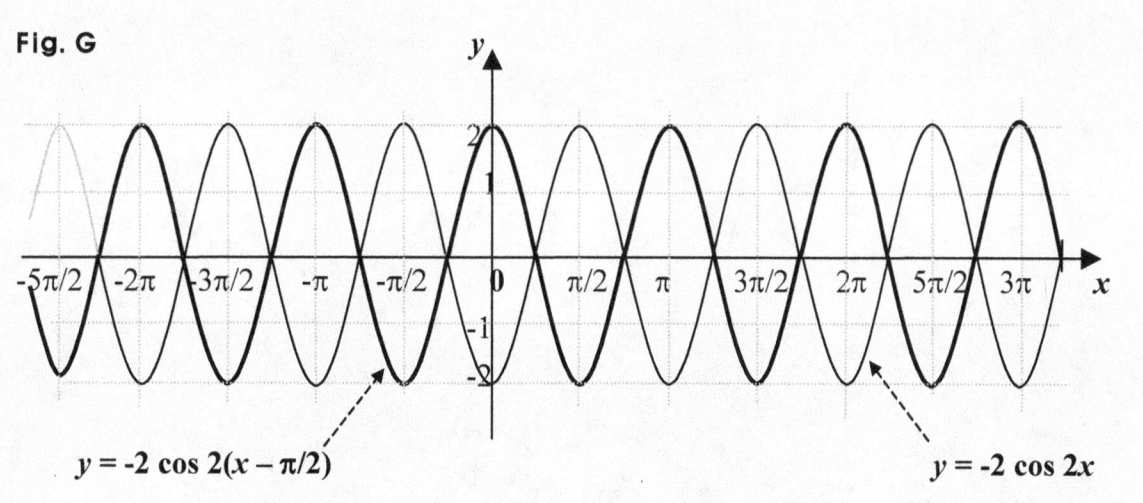

$y = -2 \cos 2(x - \pi/2)$

$y = -2 \cos 2x$

And next, setting the vertical shift to 1, we get: **-2 cos 2(x − π/2) + 1**.

Then, we move upward the curve with **−2 cos 2(x − π/2)** in the amount of 1 to get the curve with **−2 cos 2(x − π/2) + 1**, which is the curve of the function **C** given.

And thus, we can put in a graph, the curve of the function **C** the way below:

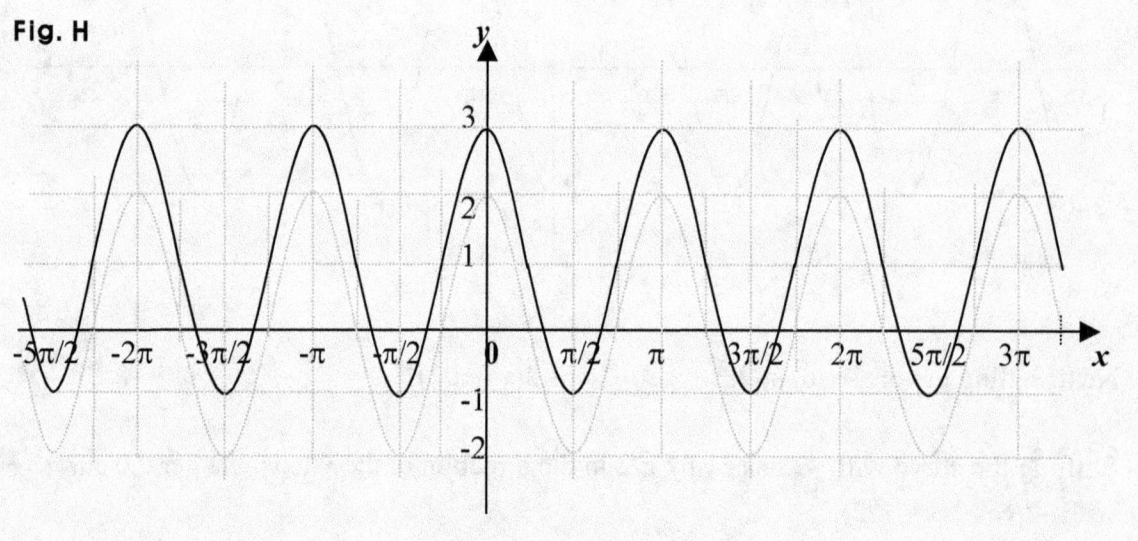

Fig. H

Examples in Cosine Functions

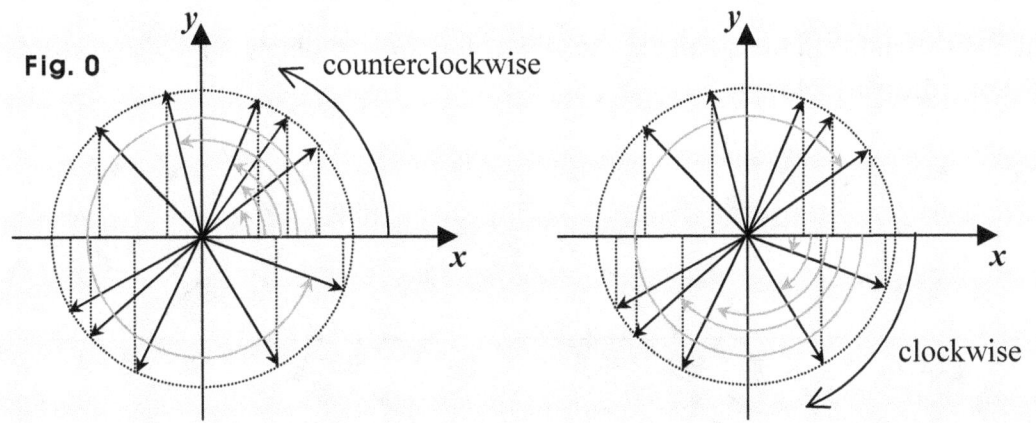

Fig. 0   counterclockwise

clockwise

In trigonometry dynamic, the ray turning makes governing angles. If turning clockwise, it makes angles negative, and if counterclockwise, it makes angles positive.

And of course, if no turning, the angle made is 0. So a governing angle can be 0° or any angle positive or negative. In each right triangle, the ray is of length 1, and is the hypotenuse. And (x, y) is the terminal point, so x is the adjacent, and y is the opposite.

And thus, assuming θ is a governing angle, we get: $\cos \theta = x$.

So if the ray is in the first quadrant, $\cos \theta > 0$, since $x > 0$.
In the second, $\cos \theta < 0$, since $x < 0$.
In the third, $\cos \theta < 0$, too, because $x < 0$. And in the fourth, $\cos \theta > 0$, since $x > 0$.

Put in a graph the curve of each of the equations below:

0. $y = \cos |2x|$ 1. $|y| = \cos 2x$ 2. $|y| = \cos |2x|$

3. $y = \cos x + \cos |x|$ 4. $y = \cos x + \sin x$

Suggestions or Solutions
To the **Problem** in the Example **0**

Put in a graph the equation as follows: $y = \cos |2x|$.

To begin with, putting in the *x-y* plane, the curve of the cosine function $y = f(x) = \cos x$, we can put it the way below:

Fig. 0.0

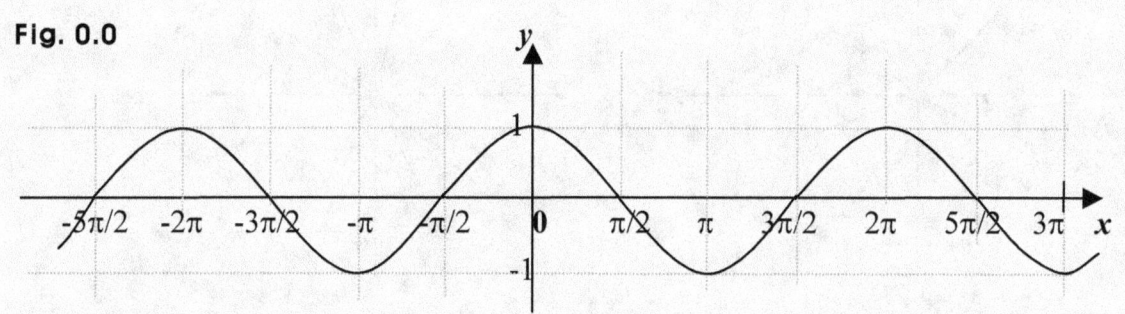

The curve above is in fact, the same as the curve of an equation $y = \cos x$.

And the curve of an equation $y = \cos 2x$ is the curve of a function $y = g(x) = \cos 2x$, too.

We know in the curve of the function **g**, the frequency is 2, so the period is $2\pi/2 = \pi$.

So putting in a graph, the curve of the equation $y = \cos 2x$, we can put it the way below:

Fig. 0.1

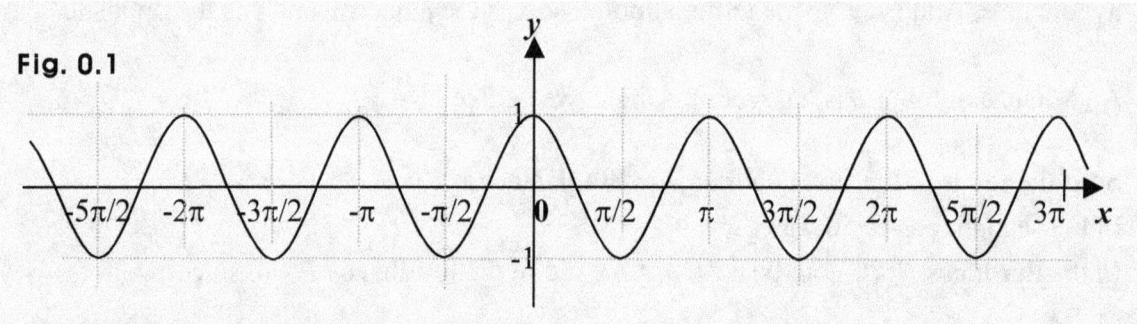

What then, about the curve of $y = \cos |x|$?

It is the same as the curve of the equation $y = \cos x$.   How come?

Though looking awkward, **cos |x|** can be put this way, too: **cos t** where **t = |x|**.

And we know: **|x| ≥ 0** for all **x**.

So even if **x < 0**, that is, **x** gets a negative value, we still get: **t > 0**, that is, **t** gets a positive value.

Thus, for instance, even if **x = -π/3**, we need to get: **cos π/3**, because **|-π/3| = π/3**.

So given **y = cos |x|** for **x ≥ 0**, we can just set: **y = cos x**.

If however, **x < 0**, we want to set: **y = cos (-x)** because **-x > 0**, since **|x| ≥ 0** in **cos |x|**.

We have a trig-identity though: **cos (-x) = cos x**.

So given **y = cos |x|**, we get: **y = cos x** for **x ≥ 0**, and for **x < 0**, too.

And thus, the curve of **y = cos |x|** is the curve of **y = cos x**, too.

So putting in a graph, the curve of **y = cos |x|**, we can put it the way below:

Fig. 0.2

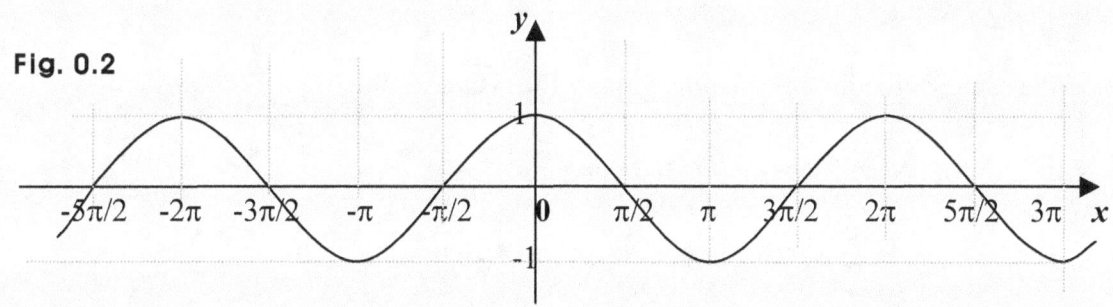

So the curve of **y = cos |x|** is the same as the curve of **y = cos x**.

And we can notice that the curve is symmetric about the **y**-axis. And in fact:
If the curve of an equation is <u>symmetric about the y-axis</u>, the equation is said to be <u>even</u>.

If the curve is <u>not symmetric</u> (asymmetric) about the **y**-axis, the equation is said to be <u>odd</u>. And the same is true for a function, too.

So the cosine function **y = c(x) = cos x** is said to be an even function since its curve is symmetric about the **y**-axis. And we know the curve of the sine function **y = s(x) = sin x** is *asymmetric* about the **y**-axis. So the sine function **y = s(x) = sin x** is an odd function.

And we can check to see if a function is even or odd the way below:

If for instance, a function $y = f(x)$ is odd, we get: $y = f(-x) = -f(x)$.

If however, $y = f(x)$ is even, we get: $y = f(-x) = f(x)$.

For instance, the cosine function $y = C(x) = \cos x$ is an even function, because we have:

$y = C(x) = \cos x = \cos(-x) = C(-x)$, that is, we have: $y = C(x) = C(-x)$.

What then, about $y = f(|x|)$?

It is even, because $y = f(|-x|) = f(|x|)$, because $|-x| = |x|$.

So for instance, if $y = f(x) = \sin x$, we get: $y = f(|x|) = \sin |x|$, and thus, we get:

$y = f(|-x|) = \sin |-x| = \sin |x|$.

And the same is true for an equation, too. So for instance:

$y = 3x^2$ is even, because we get: $y = 3(-x)^2 = 3x^2$.

$y = -5|x^3|$ is even, because we get: $y = -5|(-x)^3| = -5|-x^3| = -5|x^3|$.

$y = 2x$ is odd, because we get: $y = 2(-x) = -2x$.

$y = 2|x^3| + x$ is odd, because we get: $y = 2|(-x)^3| + (-x) = 2|x^3| - x = -(2|x^3| + x)$.

And if a function or equation is said to be even, the curve is symmetric about the y-axis.

What then, about $y = \cos 2x$?

The equation $y = \cos 2x$ is even, because we get: $\cos 2(-x) = \cos 2x$. How come?

Assuming $t = 2x$, we can set: $\cos 2x = \cos t$.

Then, we get: $2(-x) = -2x = -t$, so we get: $\cos (-2x) = \cos (-t) = \cos t = \cos 2x$.

So given $y = \cos |2x|$, we get: $y = \cos 2x$ for $x < 0$ as well as for $x \geq 0$.

Thus, the curve of $y = \cos |2x|$ is the same as the curve of $y = \cos 2x$.

And we know putting in a graph, the curve of the equation $y = \cos 2x$, we get:

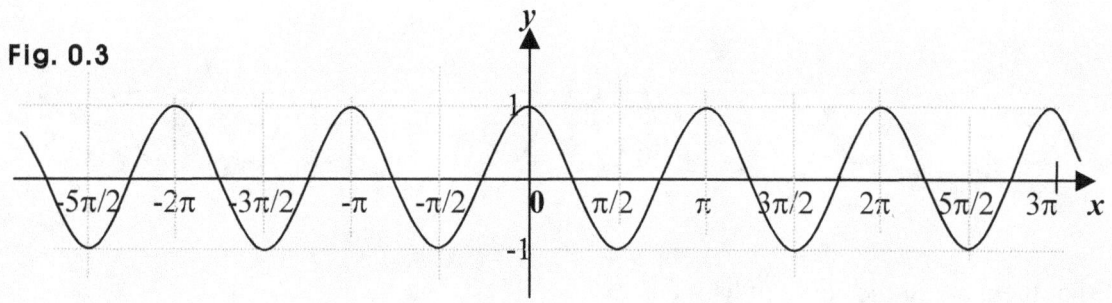

Fig. 0.3

Note that the curve of an equation or function $y = h(|x|)$ is symmetric about the y-axis.

For instance, assuming: $h(|x|) = |x|^2 - 3|x| + 2$, we get: $y = |x|^2 - 3|x| + 2$, and the curve of it is symmetric about the y-axis.

And assuming: $h(|x|) = \sin |x| + 2|x| + 1$, we get: $y = \sin |x| + 2|x| + 1$, and the curve of it is symmetric about the y-axis. And of course, not all cosine functions are even functions.

For instance, assuming: $y = g(x) = \cos (x - \pi/2)$, we get:

$g(-x) = \cos (-x - \pi/2) = \cos \{-(x + \pi/2)\} = \cos (x + \pi/2)$.

So we get: $y = g(x) \neq g(-x)$.

Suggestions or Solutions
To the **Problem** in the Example **1**

Put in a graph the equation as follows: $|y| = \cos 2x$.

To begin with, putting in the *x-y* plane, the curve of the equation $y = \cos 2x$, we can put it the way below:

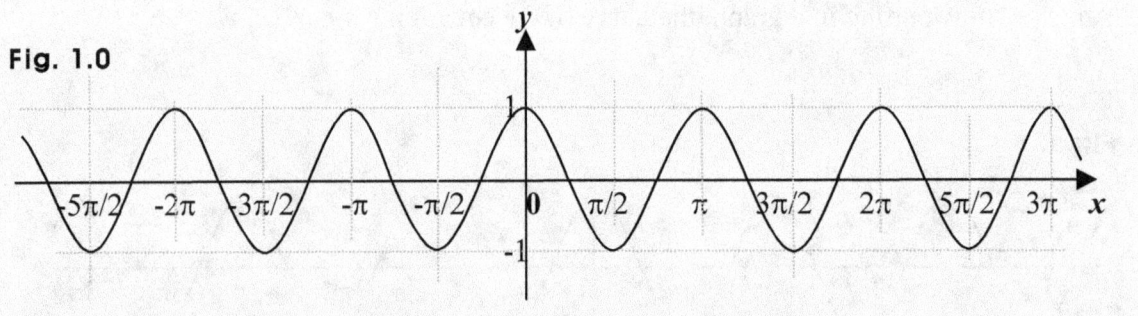

Fig. 1.0

What then, about the curve of $|y| = \cos 2x$?

In that case, we want to consider two cases, one is $y \geq 0$, and the other is $y < 0$.   Why?

Though looking awkward, $|y| = \cos 2x$ can be put this way, too: $t = \cos 2x$ where $t = |y|$.

And we know: $|y| \geq 0$ for all *y*.

So even if $y < 0$, that is, *y* gets a negative value, we still get: $t > 0$, that is, *t* gets a positive value.

So for instance, even if $y = -1/2$, we need to get: $t = 1/2$, because $|-1/2| = 1/2$.

So given $|y| = \cos 2x$ for $y \geq 0$, we can just set: $y = \cos 2x$.

If however, $y < 0$, we want to set: $-y = \cos 2x$ because $-y > 0$ since $|y| \geq 0$ in $|y| = \cos 2x$.

Thus, we get: $y = \cos 2x$ for $y \geq 0$, and we get: $y = -\cos 2x$ for $y < 0$.

And taking care of first, the curve of $y = \cos 2x$ for $y \geq 0$, we can put it this way:

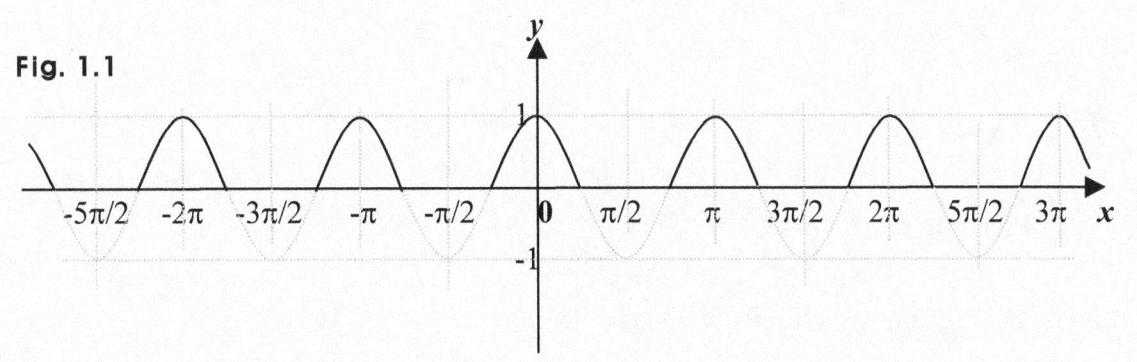

Fig. 1.1

Next, we can put in a graph, the curve of $y = -\cos 2x$ the way below:

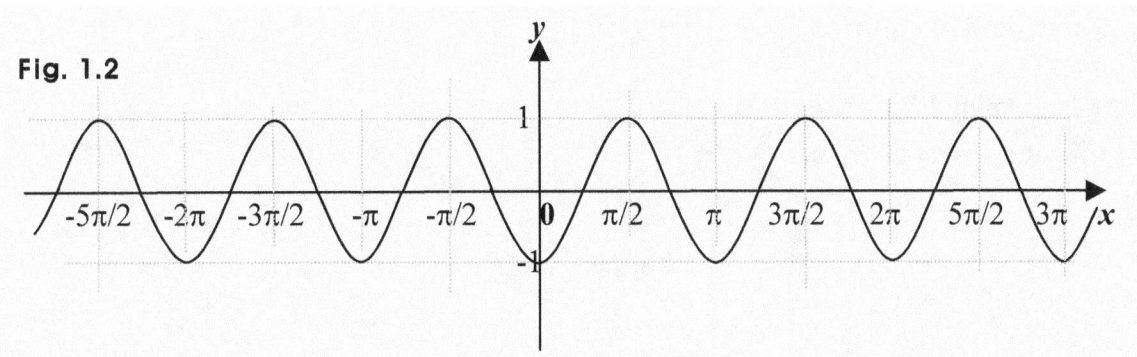

Fig. 1.2

So next, moving on to the curve of $y = -\cos 2x$ for $y < 0$, we can put it this way:

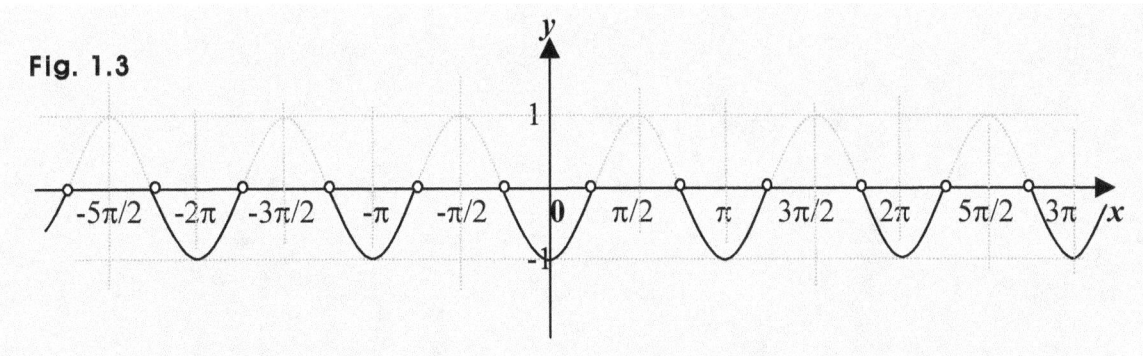

Fig. 1.3

Note that all the points $((2n + 1)\pi/4, 0)$ for n integer, do not belong to the curve above.

And thus, putting in a graph, the curve of $|y| = \cos 2x$, we can put it the way below:

Fig. 1.4

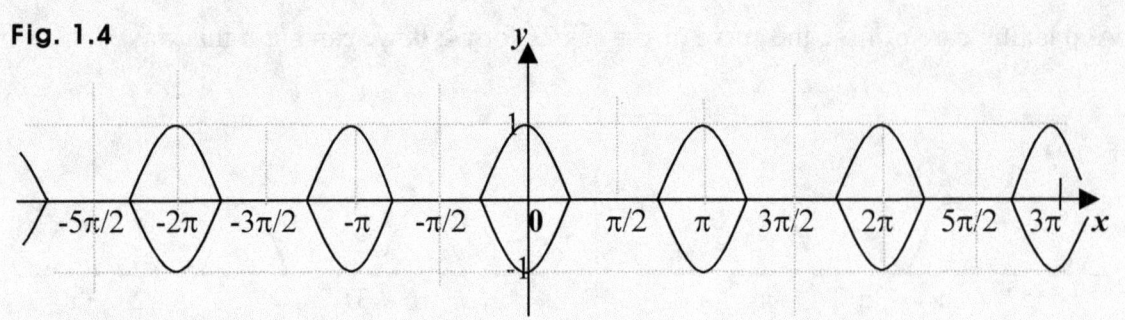

Notice that the curve is symmetric about the **x**-axis.

In fact, the curve of an equation or function $|y| = h(x)$ is symmetric about the x-axis.

For instance, assuming $h(x) = x^2 - 3x + 2$, we get: $|y| = x^2 - 3x + 2$, and the curve of it is symmetric about the **x**-axis.

And assuming $h(x) = \sin x + 2x + 1$, we get: $|y| = \sin x + 2x + 1$, and the curve of it is symmetric about the **x**-axis.

Suggestions or Solutions
To the **Problem** in the Example **2**

Put in a graph the equation as follows: $|y| = \cos |2x|$.

To begin with, putting in the *x-y* plane, the curve of the equation $y = \cos 2x$, we can put it the way below:

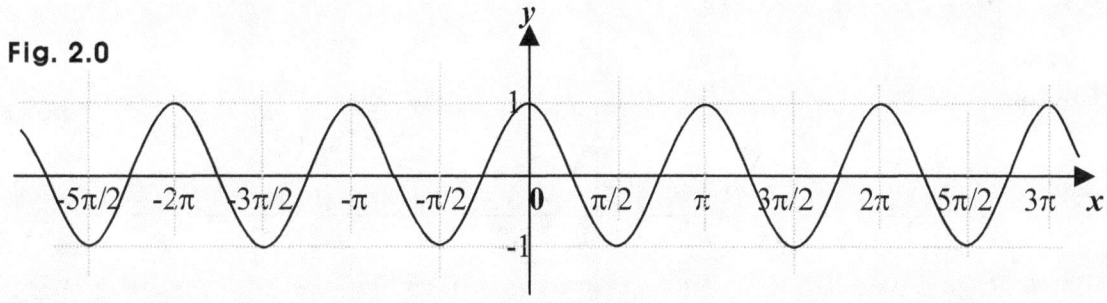

Fig. 2.0

Next, we know: $|y| \geq 0$ for all *y*, and $|2x| \geq 0$ for all *x*.

So we want to consider four different cases as follows:

First, assuming $x \geq 0$, and $y \geq 0$, we get: $y = \cos 2x$.

Next, assuming $x \geq 0$, and $y < 0$, we get: $-y = \cos 2x$, so we get: $y = -\cos 2x$.

Next, assuming $x < 0$, and $y < 0$, we get: $-y = \cos (-2x)$, so we get: $y = -\cos 2x$.

And next, assuming $x < 0$, and $y \geq 0$, we get: $y = \cos (-2x)$, so we get: $y = \cos 2x$.

Thus in sum, we have:

$y = \cos 2x$ for $x \geq 0$, and $y \geq 0$. $y = -\cos 2x$ for $x < 0$, and $y < 0$.

$y = -\cos 2x$ for $x \geq 0$, and $y < 0$. $y = \cos 2x$ for $x < 0$, and $y \geq 0$.

So beginning with the curve of $y = \cos 2x$ for $x \geq 0$, and $y \geq 0$, we get:

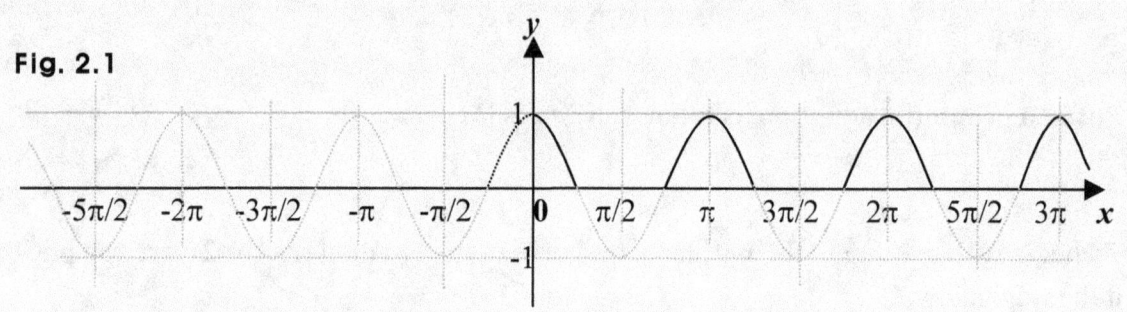

Fig. 2.1

Next, moving on to the curve of $y = \cos 2x$ for $x < 0$, and $y \geq 0$, we get:

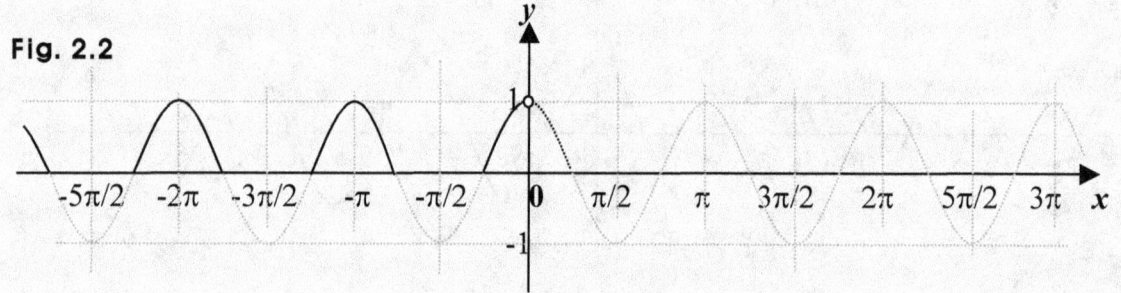

Fig. 2.2

Next, we can put in a graph, the curve of $y = -\cos 2x$ the way below:

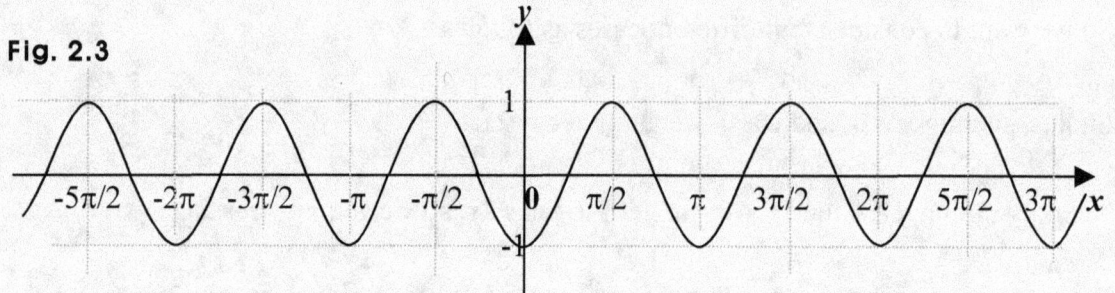

Fig. 2.3

So moving on to the curve of $y = -\cos 2x$ for $x < 0$, and $y < 0$, we get:

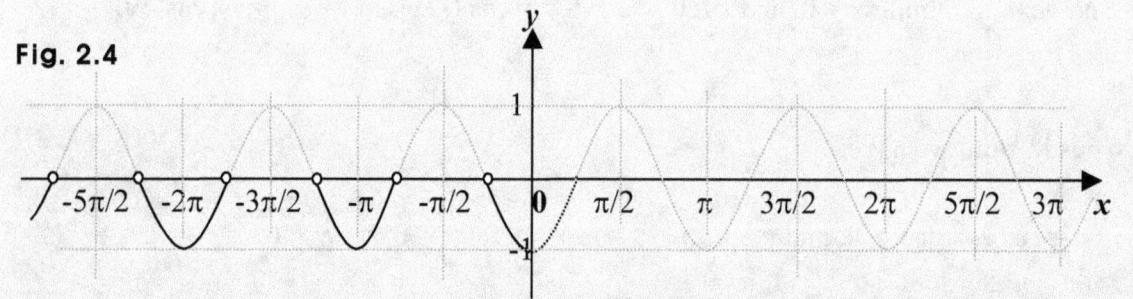

Fig. 2.4

Note that all the points $(-(2n + 1)\pi/4, 0)$ for n integer ≥ 0, do not belong to the curve.

Next, moving on to the curve of $y = -\cos 2x$ for $x \geq 0$, and $y < 0$, we get:

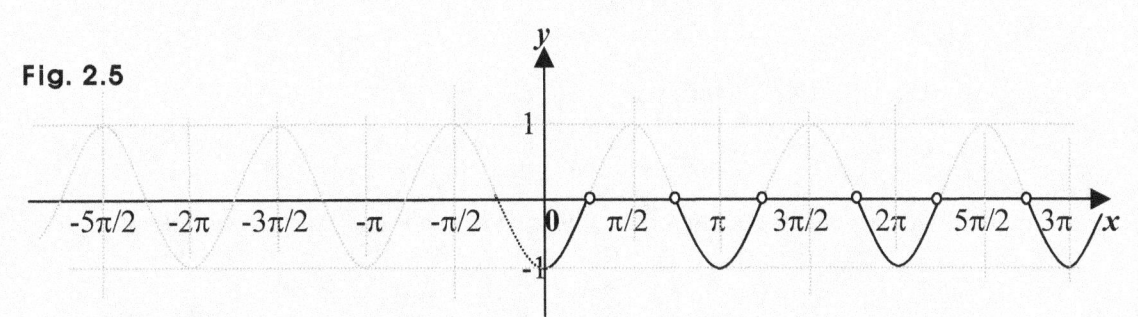

Fig. 2.5

(Note: all the points at $(2n + 1)\pi/4$ for n integer ≥ 0, do not belong to the curve above.)

And thus, putting in a graph, the curve of $|y| = \cos |2x|$, we can put it the way below:

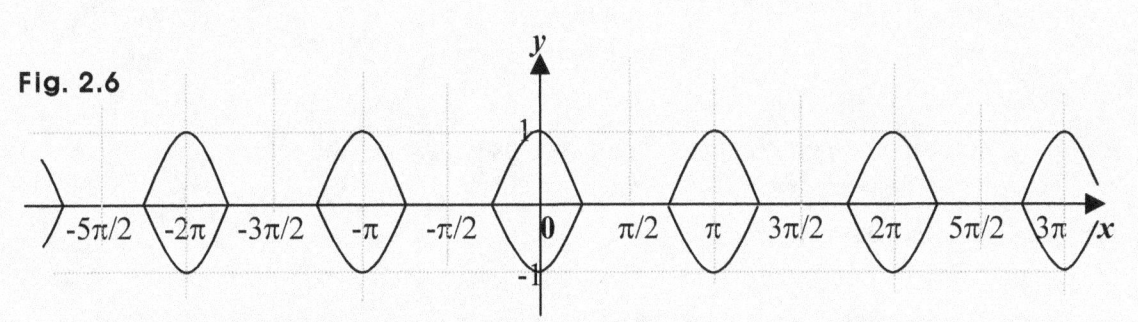

Fig. 2.6

Notice that the curve is symmetric about the origin as well as the x-axis and the y-axis.

In fact, the curve of an equation or function $|y| = h(|x|)$ is symmetric about the origin as well as the x-axis and the y-axis.

For instance, assuming $h(|x|) = |x|^2 - 3|x| + 2$, we get: $|y| = |x|^2 - 3|x| + 2$, and the curve of it is symmetric about the origin as well as the x-axis and the y-axis.

And assuming $h(|x|) = \sin |x| + 2|x| + 1$, we get: $|y| = \sin |x| + 2|x| + 1$, and the curve of it is symmetric about the origin as well as the x-axis and the y-axis.

48

Suggestions or Solutions
To the **Problem** in the Example **3**

Put in a graph the equation as follows: $y = \cos x + \cos |x|$.

To begin with, we can put in the **x-y** plane, the curve of $y = \cos x$, the way below:

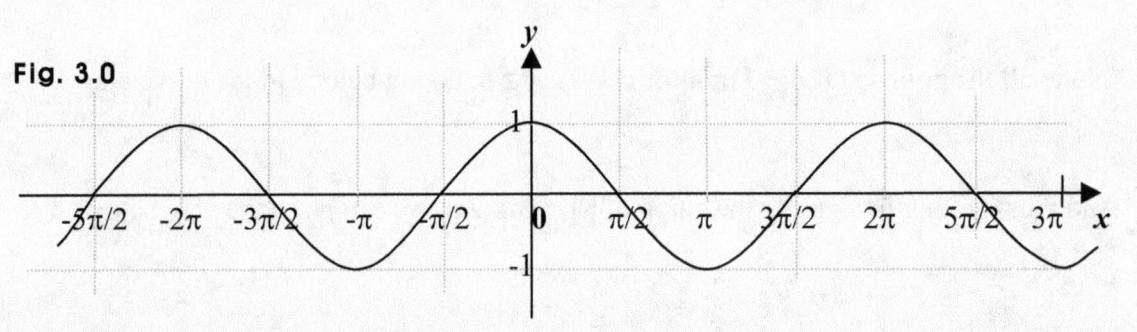

Fig. 3.0

And in fact, putting in a graph, the curve of $y = \cos |x|$, we get the same one as above.

That's because $\cos |x| = \cos x$.

So the curve of $y = \cos x + \cos |x|$ is the same as the curve of $y = 2\cos x$.

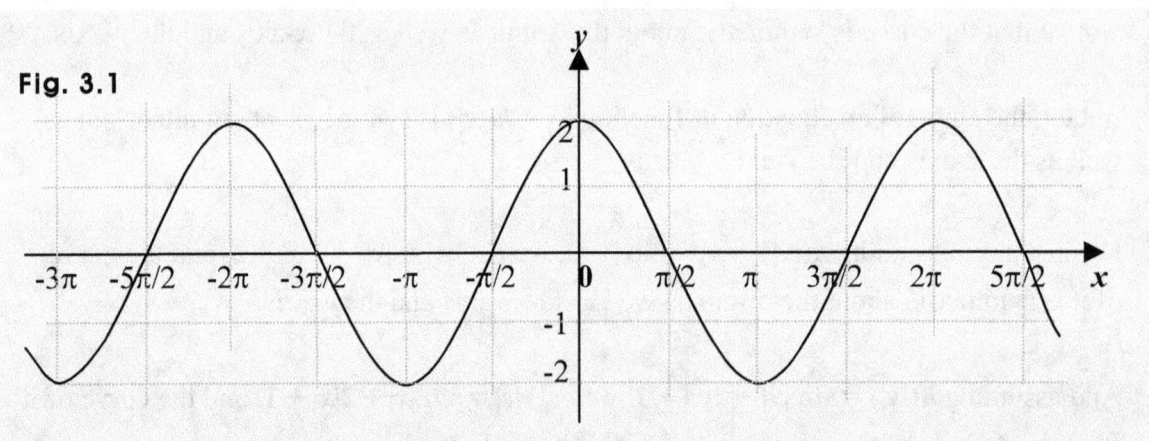

Fig. 3.1

And thus, the curve above is the curve of the equation $y = \cos x + \cos |x|$.

What then, about the curve of $y = \cos x - \cos |x|$?

We know: $\cos x = \cos |x|$.

So we get: $\cos x - \cos |x| = 0$. That is, we get: $y = 0$.

And thus, the curve of $y = \cos x - \cos |x|$ is the x-axis, itself, which is thus, as follows:

Fig. 3.2

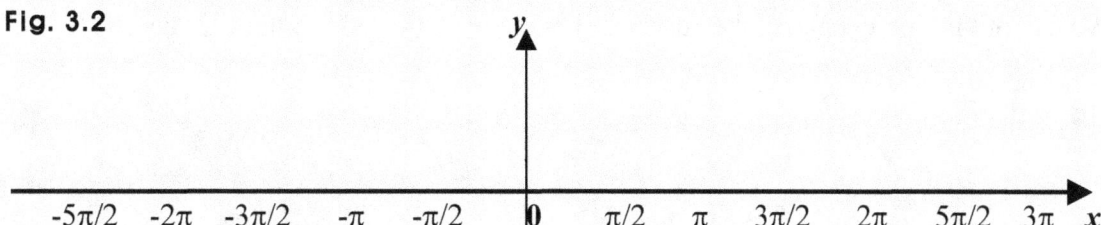

Suggestions or Solutions
To the **Problem** in the Example **4**

Put in a graph the equation as follows: $y = \cos x + \sin x$.

We can put in one *x-y* plane, the curves of $y = \cos x$ and $y = \sin x$, the way below:

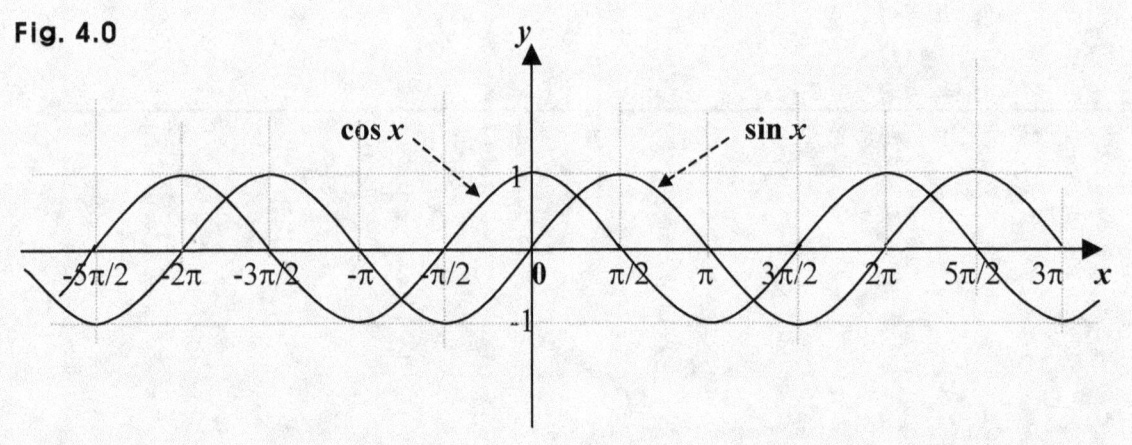

Fig. 4.0

And next, adding together the two curves above, we get another curve. And we can put in one graph, all the three curves the way below:

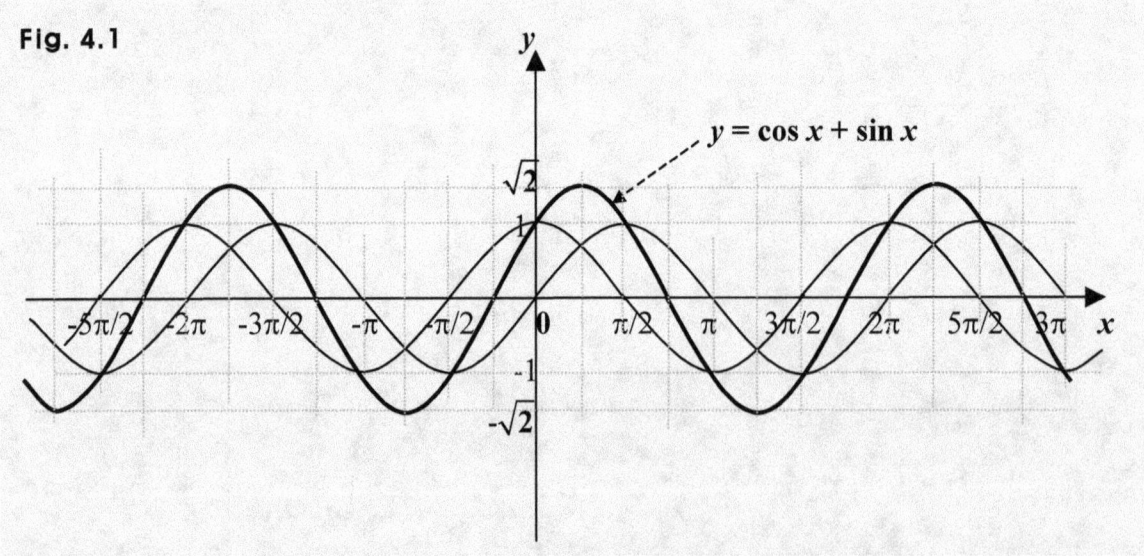

Fig. 4.1

And we can expect the curve of $y = \cos^2 x$ to be made the way below:

Fig. 4.2

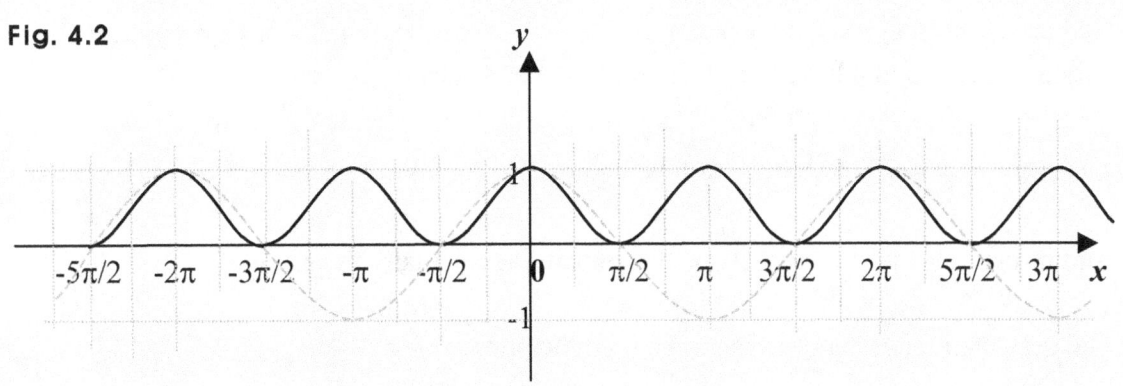

And we can put in a graph, the the curve of $y = \cos x - \sin x$ the way below:

Fig. 4.3

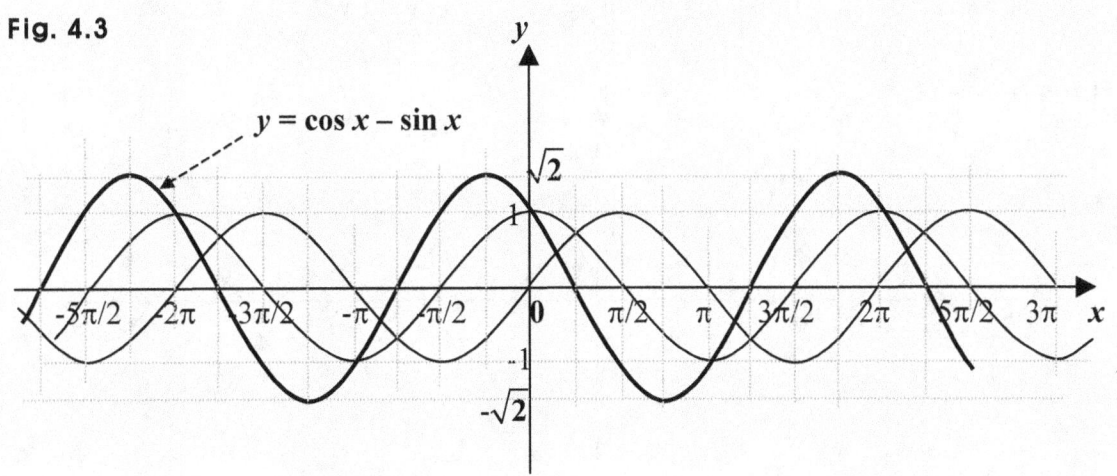

$y = \cos x - \sin x$

And also, we can put in a graph, the the curve of $y = \sin x - \cos x$ the way below:

Fig. 4.4

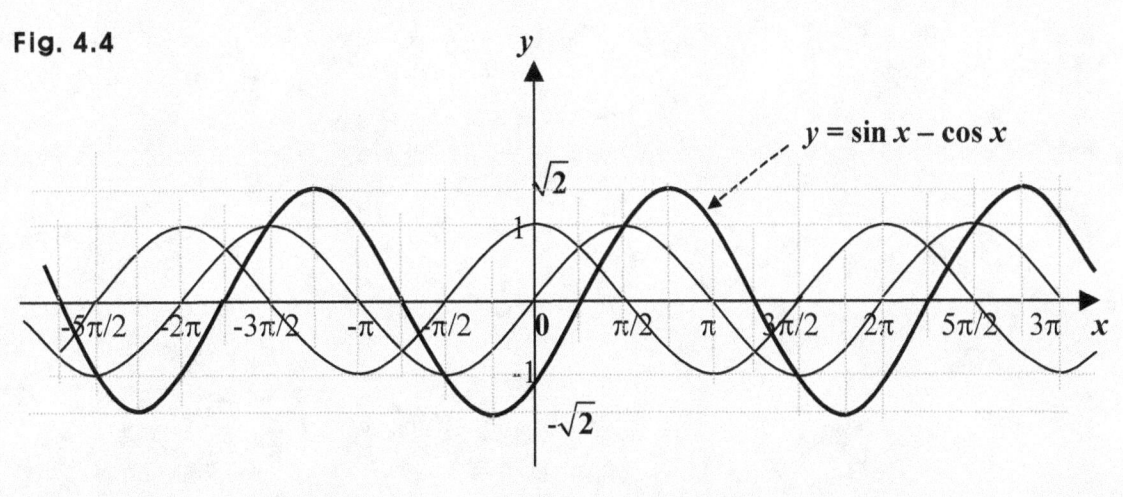

$y = \sin x - \cos x$

And assuming $f(x) = \sin x + \cos x$, $g(x) = \cos x - \sin x$, and $h(x) = \sin x - \cos x$, we can notice that f, g, and h have the same amplitude and period.

The amplitude is $\sqrt{2}$, and the period is 2π.

And in fact, the curves themselves of f, g, and h are actually the same.

The only difference between the three is in the phase.

That is to say that we have: $f(x + \pi) = g(x)$, $f(x) = g(x - \pi/2)$, and $g(x - \pi) = h(x)$.

Besides, we can put in a graph, the the curve of $y = \cos x \sin x$ the way below:

Fig. 4.5

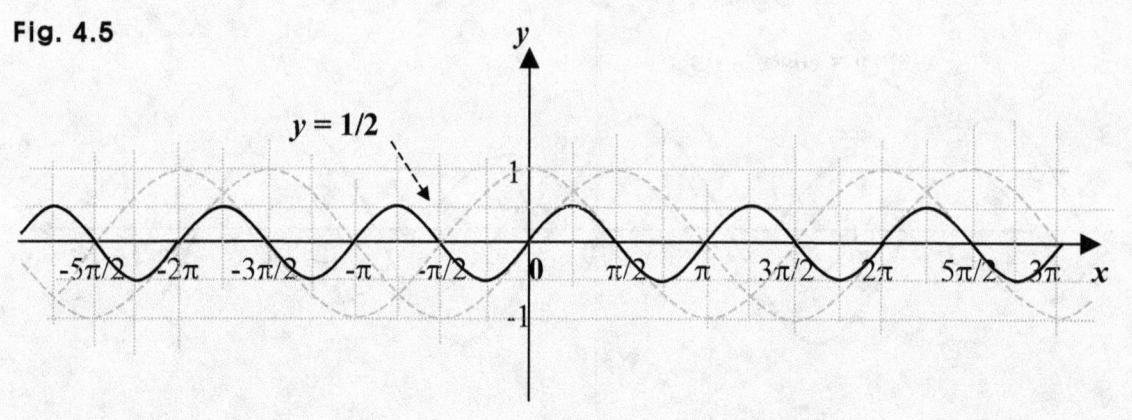

And in fact, we have: $\sin 2x = 2 \cos x \sin x$. That is, $\cos x \sin x = \frac{1}{2}\sin 2x$.

c. **Tangent Functions**

What is a tangent function?

It is a function of which each output is a trig-ratio called the tangent. So for instance, putting a cosine function in the *x-y* coordinate system, we can put it the way below:

$y = f(x) = \tan x$, where *x* is an angle.

So it is a function of an angle, and thus, takes an angle as an input.

And such an angle is a governing angle. So the input variable *x* takes a governing angle, which is an angle made by the terminal ray turning about the origin in the *x-y* plane.

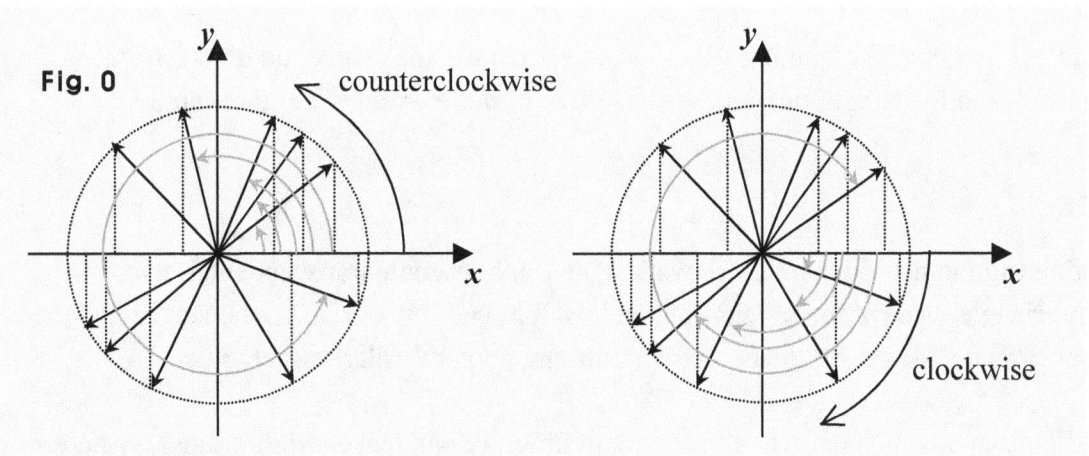

And assuming <u>the length of the ray is 1</u>, the terminal point is (*x, y*), we can see a lot of right triangles. In each, the ray is the hypotenuse, *x* is the adjacent, and *y* is the opposite.

So again, in the *x-y* plane where the ray is turning counterclockwise, placing four lamps the way below, we get shadows of the ray (the projections) on both axes.

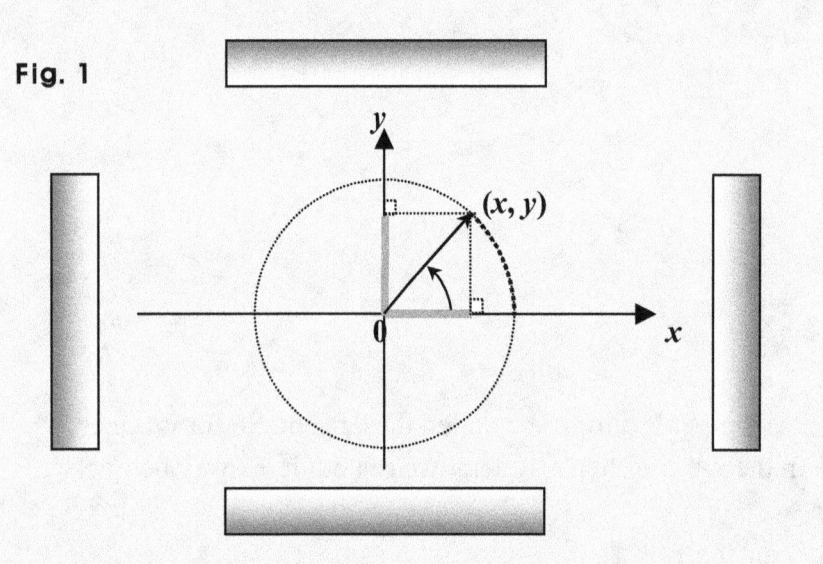

Fig. 1

Then, as the ray keeps turning, we can see on one axis, a linear motion where the shadow (projection) decreases, while on the other axis, another linear motion where the shadow increases. And such linear motions keep repeating as the ray keeps turning.

So at every moment, assuming the ray is a hypotenuse, the projection on the *x*-axis is the adjacent, and the projection on the *y*-axis is the opposite, we get a right triangle.

And in trigonometry dynamic, we work with a right triangle transcendental, which is a right triangle made of the ray turning, and two projections. One is on the *y*-axis, and the other is on the *x*-axis. So the two projections are perpendicular to each other.

And thus, at any moment while the ray is turning, we can make a right triangle, where the hypotenuse is the ray, the adjacent is the projection on the *x*-axis, that is, the *x*-coordinate at the terminal point (x, y), and the opposite is the projection on the *y*-axis, that is, the *y*-coordinate at the terminal point.

And since the length of the ray is constant, the length of the hypotenuse is constant, too, so in fact, <u>the angle made by the ray turning determines a right triangle</u>, because the angle made determines the opposite and the adjacent. And we use that angle to get the trig-ratios. So each angle made by the ray turning is a governing angle.

And when taking trig-ratios in a right triangle transcendental, too, we take the ratios the way we take them in a right triangle normal.
So it is always the case where the cosine is: the adjacent over the hypotenuse, the sine is the opposite over the hypotenuse, and the tangent is: the opposite over the adjacent.

• Now, in dynamic trigonometry, we often work with functions called trig-functions. Of each, each input is a governing angle, which is made by the ray turning in the *x-y* plane.

So in the tangent function, $y = f(x) = \tan x$, too, the input variable x gets a governing angle. Usually though, we just call governing angles, angles, and use numbers as angles, since those angles are in radian.

So for instance, assuming g is a tangent function, and is: **tan** *x*, and the domain is a set of all angles form -90° to 90°, we say that the domain is a set of all numbers from -π/2 to π/2, and can put the trig-function f this way: $y = g(x) = \tan x$ for **-π/2 ≤ x ≤ π/2**.

What then, about outputs?

They are trig-ratios, because in a tangent function $y = f(x) = \tan x$, the x gets an angle, so each value of **tan** *x*, that is, each value of $f(x)$ is a trig-ratio, which is a number.
What then, about the range?

Unlike sine or cosine functions, the range of the tangent function f above is a set of all real numbers, which are trig-ratios, of course.

So the range of f can be put this way, too: $y \in R$, where R is a set of all real numbers. How come though?

Let's now, get back to the *x-y* plane where the projections are made on both axis.

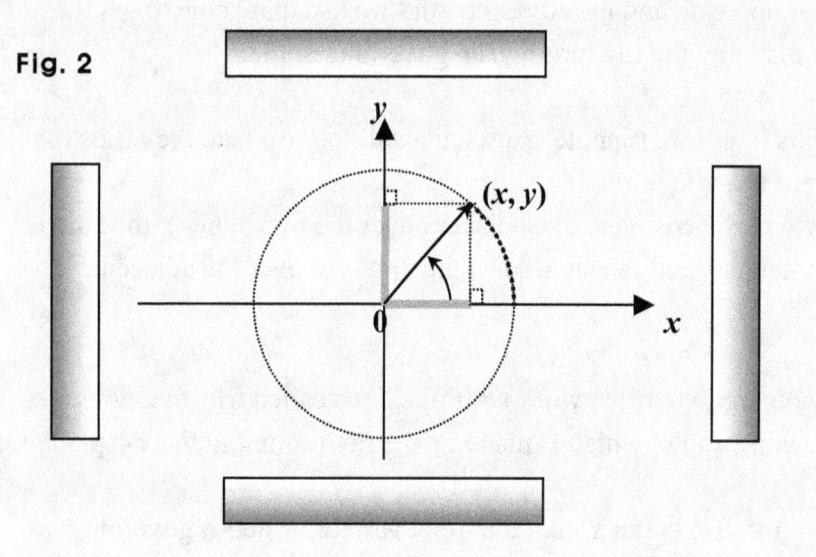

Fig. 2

Suppose that <u>the ray is of length 1</u>, and is placed on the *x*-axis to the right of the origin.

Suppose now that the ray starts turning counterclockwise.

Then, we can see that on the *x*-axis, the projection starts decreasing from 1, while on the *y*-axis, the projection starts increasing from 0.

Even before the ray turns, assuming the ray is a hypotenuse, we can get a right triangle, where the hypotenuse is the ray, the adjacent is the projection on the *x*-axis, and is the ray itself, since the ray is on the *x*-axis, and for the same reason, the opposite is 0, which is impossible though, in a right triangle normal. And thus, it can be called a right triangle transcendental.

And we know that the angle made is 0 before the ray turns, and that by definition, <u>the tangent is: the opposite over the adjacent</u>.   So we get: **tan 0 = 0/1 = 0**.

Suppose now that the angle made is θ, and that the terminal point in the ray is (***x, y***).

Then, we can get a right triangle transcendental, where the hypotenuse is the ray, the adjacent is ***x***, and the opposite is ***y***.   What then, is the tangent of the angle θ?

We know by definition, <u>the tangent is: the opposite over the adjacent</u>. And we know the opposite is y, and the adjacent is x.   So we get: **$\tan \theta = y/x$**.

So is it the case where for any angle θ, assuming that the ray is 1, and the terminal point is (x, y), we get: **$\tan \theta = y/x$**?

Unlike the sine or cosine, it is not the case. There are some angles of which the tangent cannot be defined. And those angles are as follows: $\pm\pi/2$, $\pm 3\pi/2$, $\pm 5\pi/2$, and such.

So we can notice that **$\tan \theta$** is not defined if $\theta = (2n - 1)\pi/2$ for n is an integer. That is to say that if the angle is **not** an integer multiple of $90°$, the tangent is defined.

•• So if $\theta \neq (2n - 1)\pi/2$ for n integer, the ray is 1, and the terminal point is (x, y), we get: **$\tan \theta = y/x$**, both of which are of course, the two coordinates at the terminal point (x, y).

Suppose now that the angle θ is $\pi/2$.   What right triangle then, can we get?

It is a right triangle transcendental, where the opposite is the ray, which is 1, and the adjacent is the projection on the x-axis, which is now 0.

That is, the opposite is 1, and the adjacent is 0.

So we don't get: **$\tan \pi/2$**. It's simply because the denominator is 0. And thus, the tangent cannot be defined for the angle $\pi/2$.

Suppose next that the angle θ is π.   What right triangle then, can we get?

It is a right triangle transcendental, where the opposite is 0, and the adjacent is the projection on the x-axis, which is now the ray of length 1.   So we get: **$\tan \pi = 0$**. And we know if θ is π, the ray is on the x-axis, to the left of the origin, the ray is of length 1, and the terminal point is (x, y).

So what are the *x*-value and the *y*-value at **(*x*, *y*)** if *θ* is π?

When *θ* is π, the *x*-value is -1, and is the value of the adjacent, and the *y*-value is 0, and is the value of the opposite.   So we get: **tan π = 0/(-1) = 0**.

Suppose now that the angle *θ* is 3π/2.   What right triangle then, can we get?

It is a right triangle transcendental, where the opposite is -1, and the adjacent is the projection on the *x*-axis, which is now 0. So again, we don't get: **tan 3π/2**. It's simply because the denominator is 0. And thus, the tangent cannot be defined for the angle 3π/2.

And we know if *θ* is 3π/2, the ray is on the *y*-axis, below the origin, the ray is of length 1, and the terminal point is **(*x*, *y*)**.

So what are the *x*-value and the *y*-value at **(*x*, *y*)** if *θ* is 3π/2?

When *θ* is 3π/2, the *x*-value is 0, and is the value of the adjacent, and the *y*-value is -1, and is the value of the opposite, so **tan 3π/2** is not defined, because the denominator is 0.

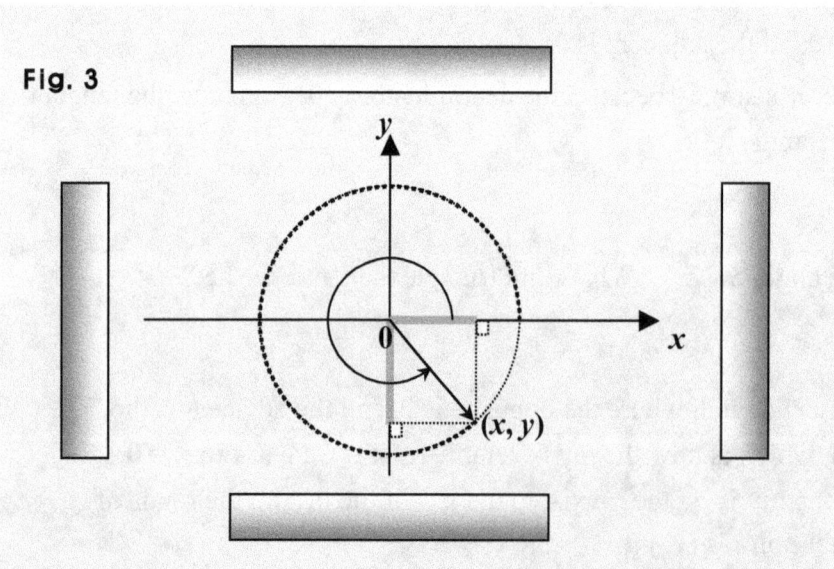

Fig. 3

So now, as the governing angle θ changes from 0 to 2π, what value does **tan** θ get?

The value of **tan** θ begins with 0, and goes to infinity as the angle θ begins with 0, and approaches $\pi/2$.   So if **$0 \le \theta < \pi/2$**, we get: $0 \le$ **tan** $\theta < \infty$, where ∞ is called infinity.

And then, the value of **tan** θ suddenly becomes negative infinity as soon as the ray passes the y-axis, and gets larger until it goes to infinity as the angle θ approaches $3\pi/2$.

So if **$\pi/2 < \theta < 3\pi/2$**, we get: $-\infty <$ **tan** $\theta < \infty$, that is, |**tan** θ| $< \infty$.

So in sum, θ cannot be $\pi/2$ and $3\pi/2$, because **tan** θ is not defined if $\theta = \pi/2$ or $3\pi/2$.

And the value of **tan** θ can be one of all real numbers as the angle θ changes from 0 to 2π skipping $\pi/2$ and $3\pi/2$.

And the same is true, too, for the case where the angle θ changes from 2π to 4π skipping $5\pi/2$ and $7\pi/2$.   That is, the value of **tan** θ can be one of all real numbers.

So we can notice that if $\theta \ne (2n-1)\pi/2$, where n is an integer ≥ 0, we get: |**tan** θ| $< \infty$. That is, the value of **tan** θ can be one of all real numbers if $\theta \ne (2n-1)\pi/2$ for n integer.

Suppose now again, the ray of length 1 is at rest on the x-axis to the right of the origin.

 • Suppose this time, the ray turns clockwise.

Then, we can see that on the x-axis, the projection starts decreasing from 1, while on the y-axis, the projection starts growing downward from 0.

Suppose now gain that the angle made is θ, and that the terminal point in the ray is (x, y).

Then, we can get a right triangle transcendental, where the hypotenuse is the ray, the adjacent is x, and the opposite is y.   What then, is the tangent of the angle θ?

We know by definition, <u>the tangent is: the opposite over the adjacent</u>. And we know the opposite is y, and the adjacent is x. So we get: **tan $\theta = y/x$**, where the angle θ is negative.

As in the case however, where the ray is turning counterclockwise, it is not the case where we get: **tan $\theta = y/x$** for any negative angle θ.

The tangent cannot be defined if the angle is: $\pm\pi/2$, $\pm3\pi/2$, $\pm5\pi/2$, or such.
So **tan θ** is not defined if $\theta = (2n-1)\pi/2$ for n integer.
In other words, if the angle is **<u>not</u> an integer multiple of $90°$**, the tangent can be defined.

•• So if $\theta \neq (2n-1)\pi/2$ for n integer, the ray is 1, and the terminal point is (x, y), we get: **tan $\theta = y/x$**, both of which are of course, the two coordinates at the terminal point (x, y).

Suppose now that the angle θ is $-\pi/2$.   What right triangle then, can we get?

It is a right triangle transcendental, where the opposite is the ray of length 1, and thus, is -1, since the ray is on the y-axis, and is below the origin, and the adjacent is the projection on the x-axis, which is now 0, which cannot be however, a denominator. So we don't get: **tan ($-\pi/2$)**. That is, the tangent cannot be defined for the angle $-\pi/2$.

Suppose next that the angle θ is $-\pi$.   What right triangle then, can we get?

It is a right triangle transcendental, where the opposite is the projection on the y-axis, which is now 0, and the adjacent is the ray of length 1, and thus, is -1, since the ray is on the x-axis, and is to the left of the origin.   So we get: **tan ($-\pi$) = 0/(-1) = 0**.

And we know if θ is $-\pi$, the ray is on the x-axis, to the left of the origin, the ray is of length 1, and the terminal point is (x, y).
So what are the x-value and the y-value at (x, y) if θ is $-\pi$?

When θ is -π, the *x*-value is -1, and thus, is the value of the adjacent, and the *y*-value is 0, and thus, is the value of the opposite.

Suppose now that the angle θ is -3π/2.   What right triangle then, can we get?

It is a right triangle transcendental, where the opposite is the ray of length 1, and thus, is 1 since the ray is on the *y*-axis, and is above the origin, and the adjacent is the projection on the *x*-axis, which is now 0.

So we don't get: **tan (-3π/2)** since a denominator cannot be 0. And thus, the tangent cannot be defined for the angle -3π/2.

•• So if $\theta \neq$ **(2n − 1)π/2** for *n* integer, the ray is 1, and the terminal point is **(x, y)**, we get: **tan θ = y/x**, both of which are of course, the two coordinates at the terminal point **(x, y)**.

••• So using the facts above, we can form a function of the angle θ.   How?

As the angle θ changes, the tangent trig-ratio **tan θ** changes. So we can take as an input, each angle that θ gets, and take as the output, each trig-ratio that **tan θ** produces.

Assuming thus, the function is *f*, the domain is a set of all angles less **(2n + 1)π/2** for *n* integer, and *t* is the output variable, we can set: *t* = *f*(θ) = **tan θ**. And we call the function *f* a tangent function.   What then, about the curve?

Assuming the ray is 1, and the terminal point is **(x, y)**, we get: **tan θ = y/x**. So <u>the value of **tan θ** is: the *y*-value over the *x*-value at the terminal point **(x, y)**</u> in the ray that makes the angle θ.

So let's now get back to the *x-y* plane where the ray is turning, and see how we can get the curve of the tangent function *f*.

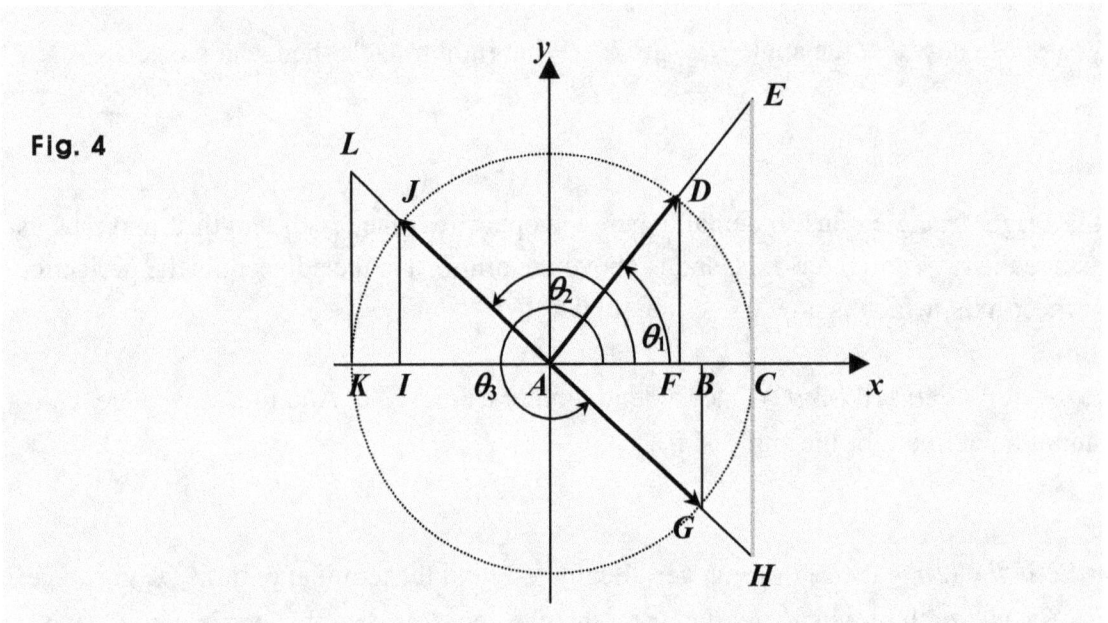

Fig. 4

First, we know: $AC = 1$, so we can get: $\tan \theta_1 = FD/AF = CE/AC = CE/1 = CE$.

Next, we can get: $\tan \theta_2 = IJ/AI = KL/AK = KL/(-1) = -KL$, which is negative.

Note that the opposite **> 0**, but the adjacent **< 0** if the ray is in the second quadrant.

So if the ray is in the second quadrant, the tangent is negative.
If however, the ray is in the third quadrant, the tangent is positive, because both the opposite and the adjacent are negative.

Next, assuming *LH* is a line segment, we can see that $CH = -KL$, because the two triangles *AKL* and *ACH* themselves are the same, and *CH* is below the *x*-axis.

So next, we can get: $\tan \theta_3 = BG/AB = CH/AC = CH/1 = CH$.

And thus, we can get the curve of $\tan \theta$ the way as follows:

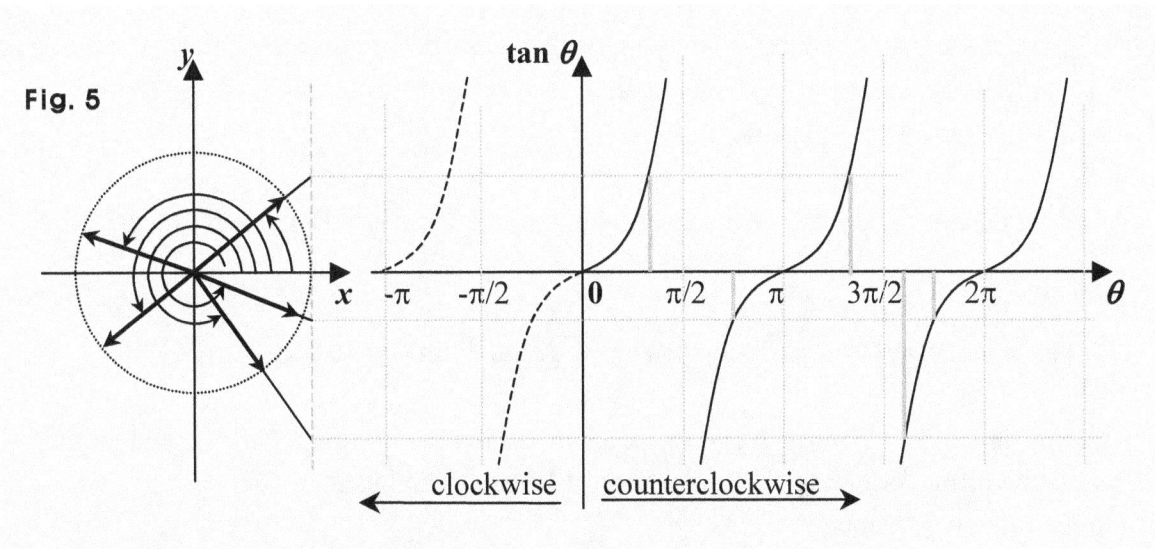

Fig. 5

And of course, we can put it in the *x-y* system.

Simply replacing *c* with *y*, and θ with *x*, we get: $y = f(x) = \tan x$.

And we can use numbers as angles. So assuming the domain is a set of all angles less **(2n + 1)90°** for **n** integer, we can set: $y = f(x) = \tan x$ for $x \neq (2n + 1)\pi/2$ for **n** integer.

And also, just setting: $y = f(x) = \tan x$, we mean that the domain is a set of all real numbers less **(2n + 1)π/2** for **n** integer.

And we can put in the *x-y* plane, the curve of the tangent function above the way below:

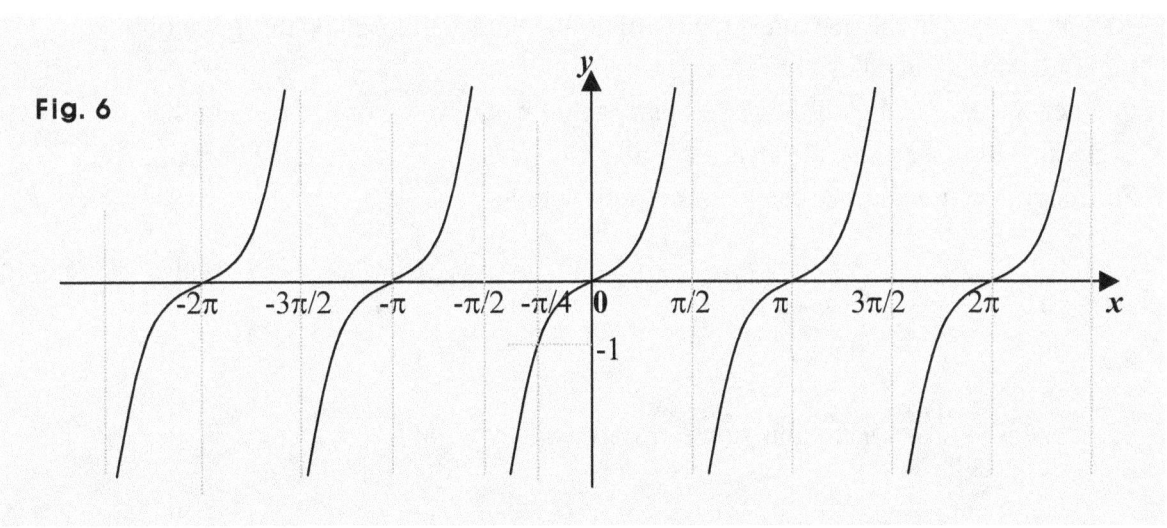

Fig. 6

And we can see that the output is not bounded, and can be any real number. So the range is a set of all real numbers. And of course, even if the domain is: $-\pi/2 < x < \pi/2$, the range is a set of all real numbers, too.

And of course, if the domain is not a set of all real numbers less $(2n + 1)\pi/2$ for n integer, the range can be other than a set of all real numbers.

If for instance, $y = g(x) = \tan x$ for $-\pi/4 \leq x < \pi/2$, the range is: $y \geq -1$.

And the tangent function $y = f(x) = \tan x$ can be called the prototype, and is thus, in the most basic form. Assuming F is a tangent function, too, and using a general form, we can put it the way below:

$y = F(x) = A \cdot \tan \{w(x + a)\} + b$, where A, w, a, and b are constant.

Or just simply, we often put it this way, too: $y = F(x) = A \cdot \tan \{w(x + a)\} + b$.

What then, about the domain?

We know that the tangent is **not** defined if the angle is $(2n + 1)\pi/2$ for n integer.
And also, we know that the angle in the function F is $w(x + a)$, and not just x.

So we want to get: $w(x + a) \neq (2n + 1)\pi/2$ for n integer.
And we know that the domain is the set of all the values that can be the value of the input variable, which is x in this case.
In other words, finding all the values that cannot be the value of x, first, and then, excluding all those values from a set of all real numbers, we can get the domain.
So finding the domain, we can get it the way below:

$$w(x+a) = \frac{(2n+1)\pi}{2} \Rightarrow x + a = \frac{(2n+1)\pi}{2w} \Rightarrow x = \frac{(2n+1)\pi}{2w} - a.$$

So if $x \neq \dfrac{(2n+1)\pi}{2w} - a$, the function F is defined.

And thus, the domain is a set of all angles less $\dfrac{(2n+1)\pi}{2w} - a$, where n is an integer, that

is, a set of all real numbers other than $\dfrac{(2n+1)\pi}{2w} - a$, where n is an integer.

And $|w|$ is called the frequency, $\frac{\pi}{|w|}$ is the period, and a is called the phase.

So in the function $y = f(x) = \tan x$, the phase is 0, and the frequency is 1, so the period is π (i.e., 180°).   How come though?

The prototype in tangent functions can be put this way: $y = f(x) = \tan x$.

And we can put the prototype into the general form, the way below:

$y = f(x) = 1 \cdot \tan 1(x + 0) + 0$.

So the phase is 0, and the frequency is 1, so the period is π.
What do we mean by the period?

Putting in the *x-y* plane, the curve of the tangent function *f* above, we get:

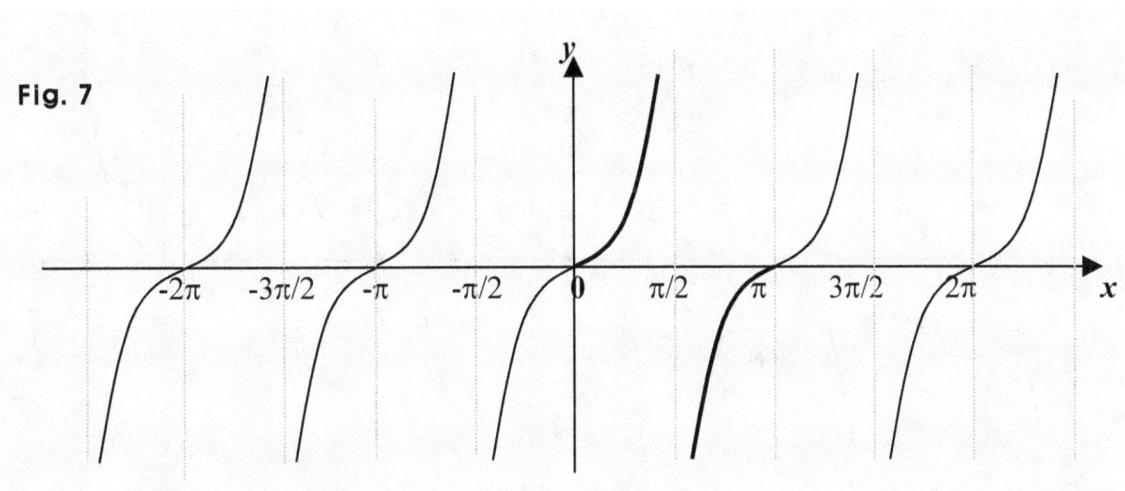

Fig. 7

Then, we can see that a part of the cuve repleats itself, which is in two pieces though.

And the part is in fact, the smallest part repeating itself.   What then, is the part?

It is the part from 0 to π. So the part repeats itself every π interval.

And we call such an interval a period.

So the interval π is the period in the function $y = f(x) = \tan x$.

• And thus, we call a tangent function a *periodic* function. And of course, the same is true, too, for sine and cosine functions, and other trig-functions as cotangent functions.

The tangent function f above is the prototype, and thus, is in the most basic form.

   • So the period π can be called the *basic period* in *tangent functions*. In other words:

   • The interval π can be called the *basic interval* in *tangent functions*.

And for instance, in the general form, setting w to 2, a and b to 0 each, and A to 1, we get a new function where: $y = g(x) = \tan 2x$ where $2x \neq (2n + 1)\pi/2$ for n integer.

Then, putting in the x-y plane, the curve of the tangent function g above, we get:

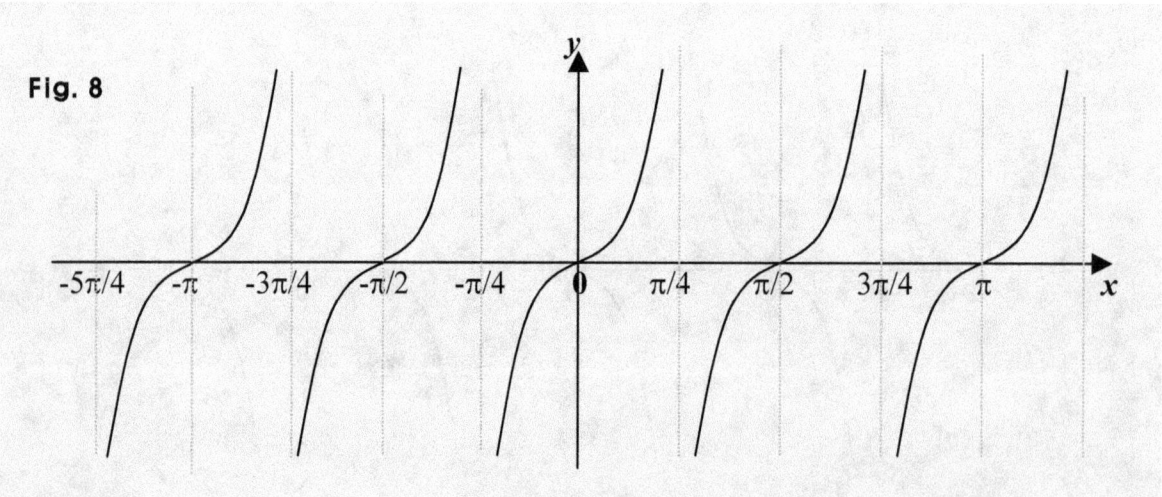

Fig. 8

Then again, we can see that a part of the cuve repleats itself.

And the part is of course, the smallest part repeating itself.   What then, is the part?

It is the part from 0 to $\pi/2$.   So the part repeats itself every $\pi/2$ interval.

And we call such an interval a period. So the interval $\pi/2$ is the period in the function *g*. How many times does the smallest part stated above appear in the basic interval π?

It appears twice in the basic interval π, which is the basic period. Then, we say the frequency is 2 in the function *g*.   What then, do we mean by the frequency?

It is the number of times the the period appears in the basic period π, which is the basic interval. In short, the frequency is the number of times the period appears in π.

The basic period however, in sine and cosine functions is not π but 2π. So in general, the frequency in a trig-function is the number of times its period appears in its basic period.

And thus, in the general form $A \cdot \tan w(x + a) + b$, $|w|$ is the frequency.

Then, given a tangent function in the general form, how can we get its period?

We know that $|w|$ is the frequency, that is, the number of times the period appears in the basic period π. So using the frequency $|w|$ and the basic period π, we can express the period this way: $\frac{\pi}{|w|}$.   Why not just *w* but $|w|$ though?

We can have a tangent function where $y = h(x) = \tan(-x)$ for *x* real. Then, the frequency in the tangent function *h* is $|-1|$, simply because a frequency is positive.

And we have: **tan (-x) = -tan x**. So assuming the ray (hypotenuse) is of length 1, we get:

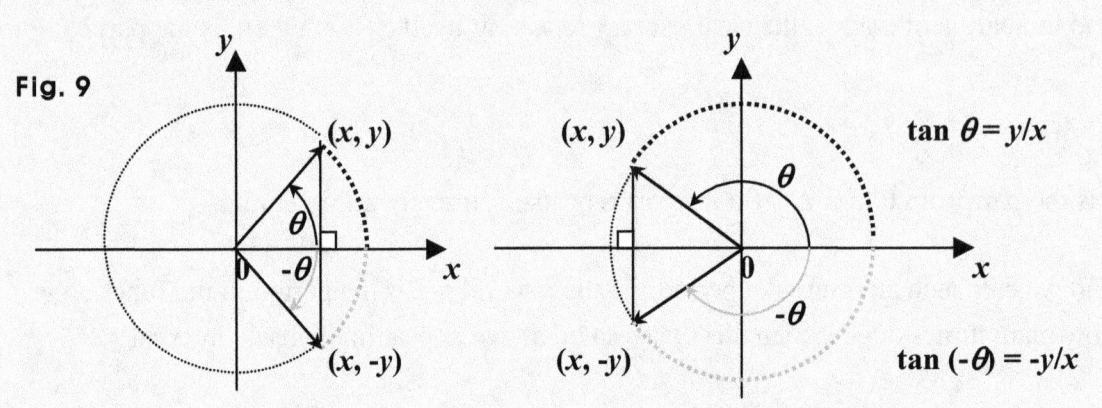

Fig. 9

So we get: **tan (-θ) = -tan θ.**   And putting in the **x-y** plane, the curve of **h**, we get:

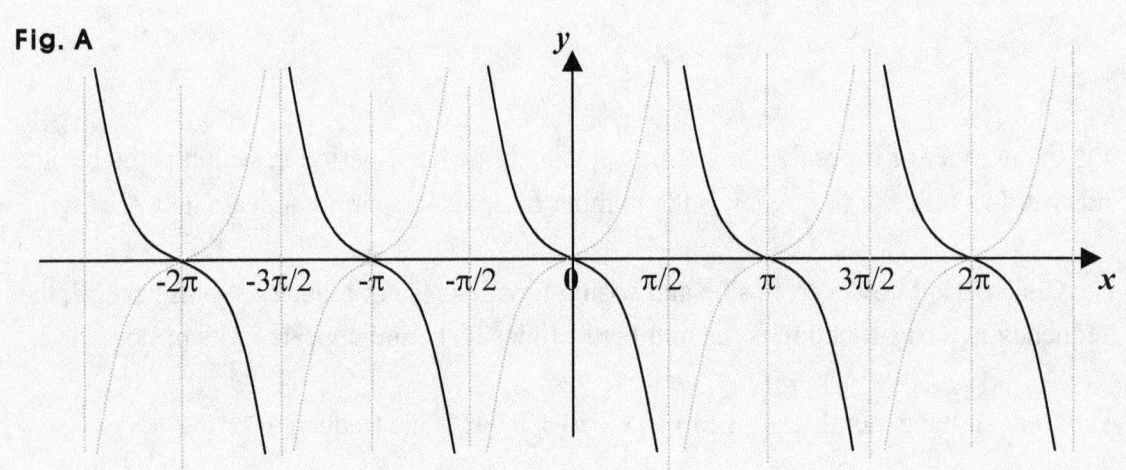

Fig. A

• And next, what do we mean by the phase?

Using the general form, we can put a tangent function called **F** the way below:

y = F(x) = A·tan w(x + a) + b where **w(x + a) ≠ (2n + 1)π/2** for **n** integer.

Then, **a** is called the phase, **|w|** is called the frequency, and $\frac{\pi}{|w|}$ is the period.

Next, shifting by **−a** in the direction of the **x**-axis, the curve of **y = Q(x) = A·tan wx + b** where **wx ≠ (2n + 1)π/2** for **n** integer, we get the curve of the function **F**.

So the two curves themselves of **F** and **Q** are the same, and moving the curve of **Q** in the amount of *–a* along the *x*-axis, we get the curve of **F**.

And thus, as in the case of sine or cosine functions, if the phase *a* is positive, the curve gets shifted to the left, and if negative, it gets shifted to the right.

Assuming for instance, shifting by $\pi/2$ in the direction of the *x*-axis, that is, to the right, the curve of *p(x)* = **tan** *x* for **-π/2 < x < π/2**, we get the curve of a function below:

q(x) = **tan** (*x* – *π/2*) for **0 < x < π**.   And putting the two curves in a graph, we get:

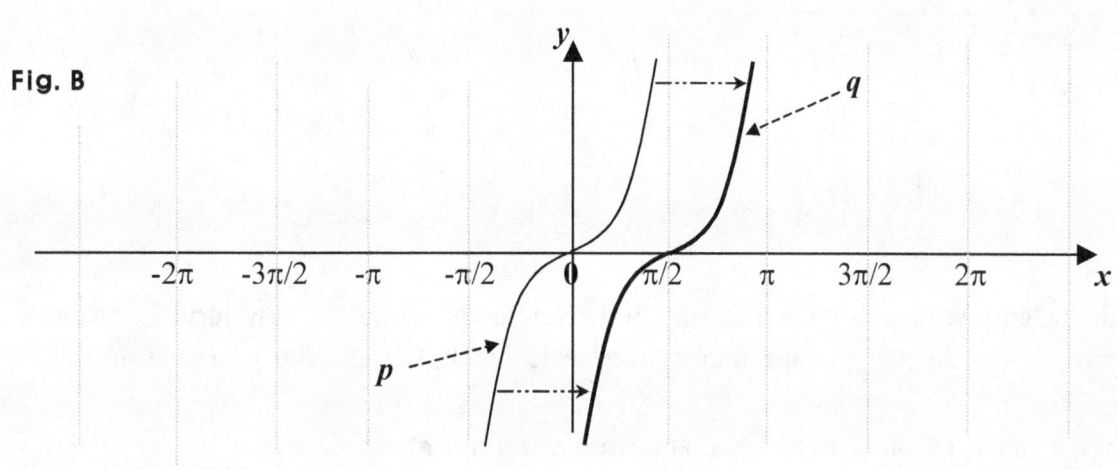

Fig. B

So the curve of *p* gets moved to the right in the amount of $\pi/2$. And the new curve is the curve of *q*. So we can call the phase a horizontal shift, too.

What then, about the constant *b* in *y* = *Q(x)* = *A*·**tan** *wx* + *b*?

It is called a vertical shift. So shifting the curve of *y* = *V(x)* = *A*·**tan** *w*(*x* + *a*) by *b* in the direction of the *y*-axis, we get the curve of the function **F** above.

So the two curves themselves of **F** and **V** are the same, and moving the curve of **V** in the amount of *b* along the *y*-axis, we get the curve of **F**.

And if *b* is positive, the curve gets shifted upward, and if negative, it moves downward.

So assuming for instance, shifting the curve of $u(x) = \tan x$ by 1 in the direction of the y-axis, that is, upward, we get the curve of a function as follows: $v(x) = \tan x + 1$. And putting u and v in a graph, we get:

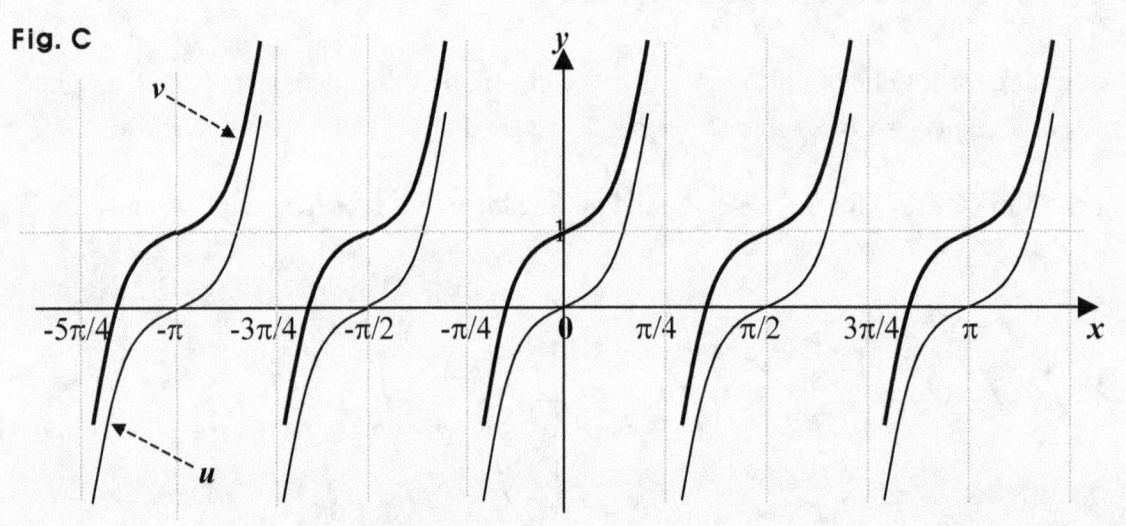

Fig. C

So the curve of u gets moved upward in the amount of 1. And the new curve is the curve of v. So we can call the b the amount of a vertical shift. In short, b is a vertical shift.

What about A though, in the general form $A \cdot \tan w(x + a) + b$?

We know A represents a number, and in fact, is a constant.

So for instance, assuming $y = B(x) = \tan x$, and multiplying by A the y-value in B for an x-value $\pi/6$, we get the y-value in a function $y = D(x) = A \cdot \tan x$ for the same x-value $\pi/6$.

That is to say that we get: $D(x) = A \cdot B(x)$.

If for instance, $A = 2$, and $x = \pi/4$, we get: $D(\pi/4) = 2B(\pi/4) = 2$ since $\tan \pi/4 = 1$.

And let's now put in a graph, for instance, the curve of a tangent function below:

$y = T(x) = -2 \tan (-2x + \pi/2) + 1$ where $-2x + \pi/2 \neq (2n + 1)\pi/2$ for n integer.

Then. to begin with, we have: **tan (-θ) = -tan θ.**

So we can get: **tan (-2x + π/2) = tan {-(2x – π/2)} = -tan (2x – π/2).**

Thus, we get: **-2 tan (-2x + π/2) + 1 = -2 tan (2x – π/2) + 1.**

And putting it in the general form, we get: **-2 tan 2(x – π/4) + 1.**

So we can now see that the frequency is 2, the phase is –π/4, that is, shifting by π/4 to the right, and the vertical shift is 1.

And next, putting it in a graph, we may want to begin with the prototype: **tan x.**

First, setting the frequency to 2, we get: **tan 2x.**

We know that the period is: the basic period over the frequency, and that the basic period for the tangent is π. So we can see that the period is: π/2.

And thus, we can put the curve with **tan 2x** the way below:

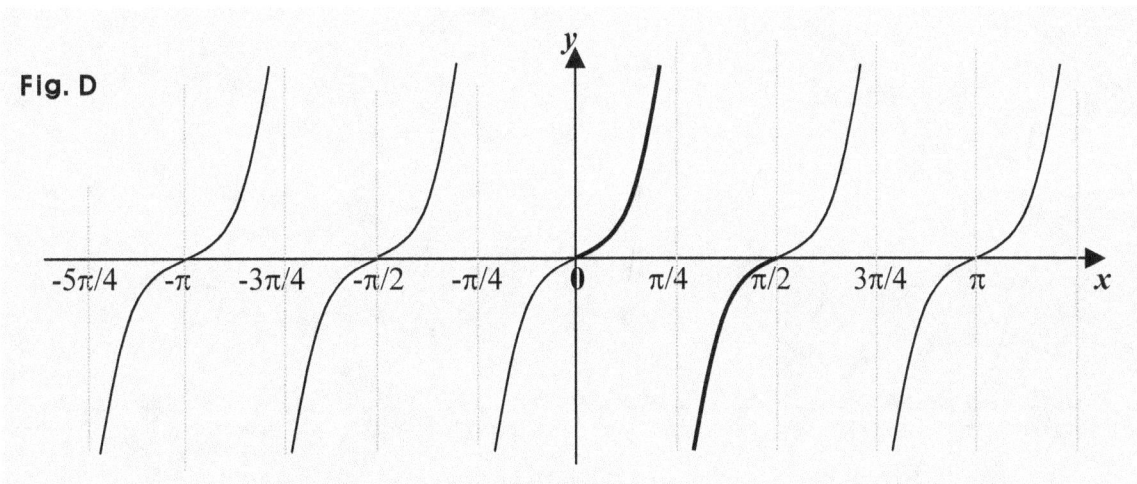

Fig. D

Next, setting the constant A to 2, we get: **2tan 2x.**

And assuming $m(x)$ = **tan 2x** and $n(x)$ = **2tan 2x**, we get: $n(x)$ = 2$m(x)$ for every x-value.

And thus, we can put in a graph the curve with **2tan 2x** the way below:

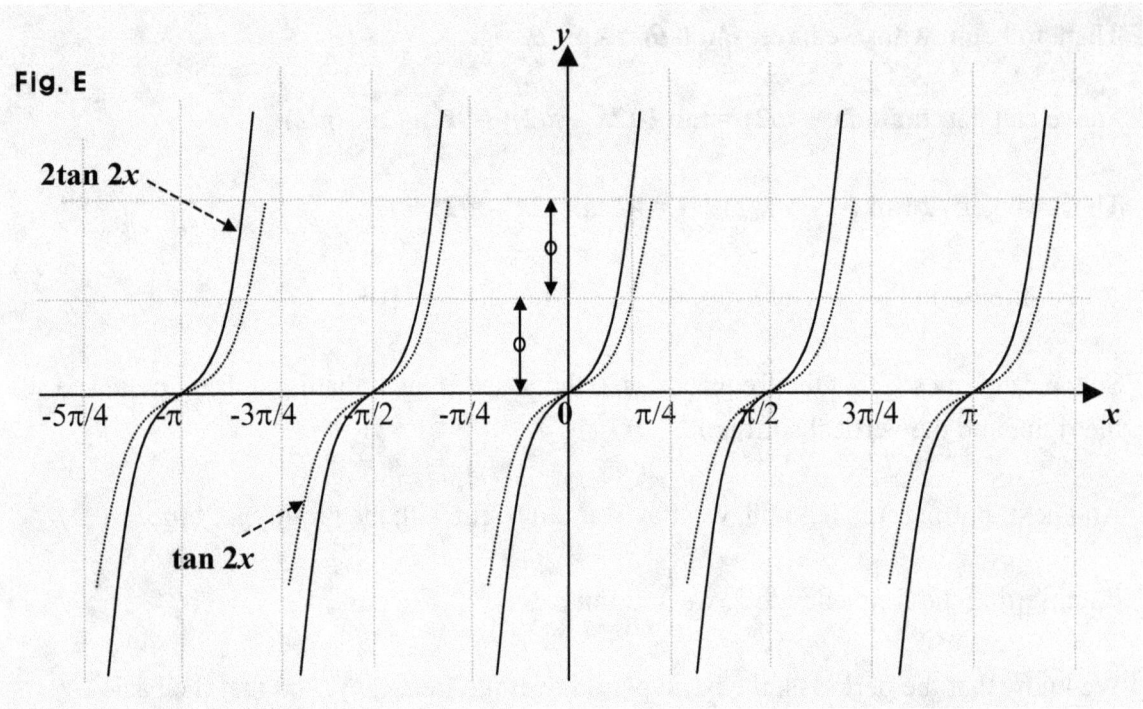

And we want to put a negative sign in front.   So we get: **-2tan 2***x*.

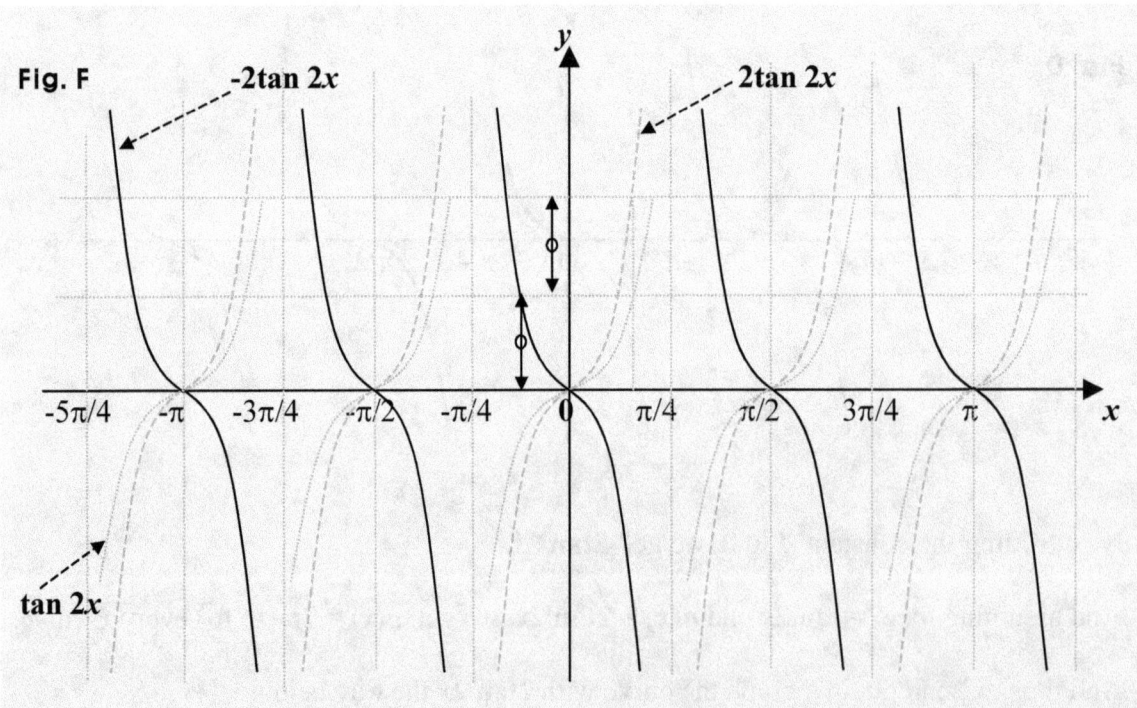

Next, setting the phase to $-\pi/4$, we get: **-2tan 2(x − π/4)**.

Shifting the curve with **-2tan 2x** by $\pi/4$ in the direction of the **x**-axis, we get the curve with **-2tan 2(x − π/4)**.

So moving the curve with **-2tan 2x** in the amount of $\pi/4$ to the right, we get the curve with **-2tan 2(x − π/4)**, which is below:

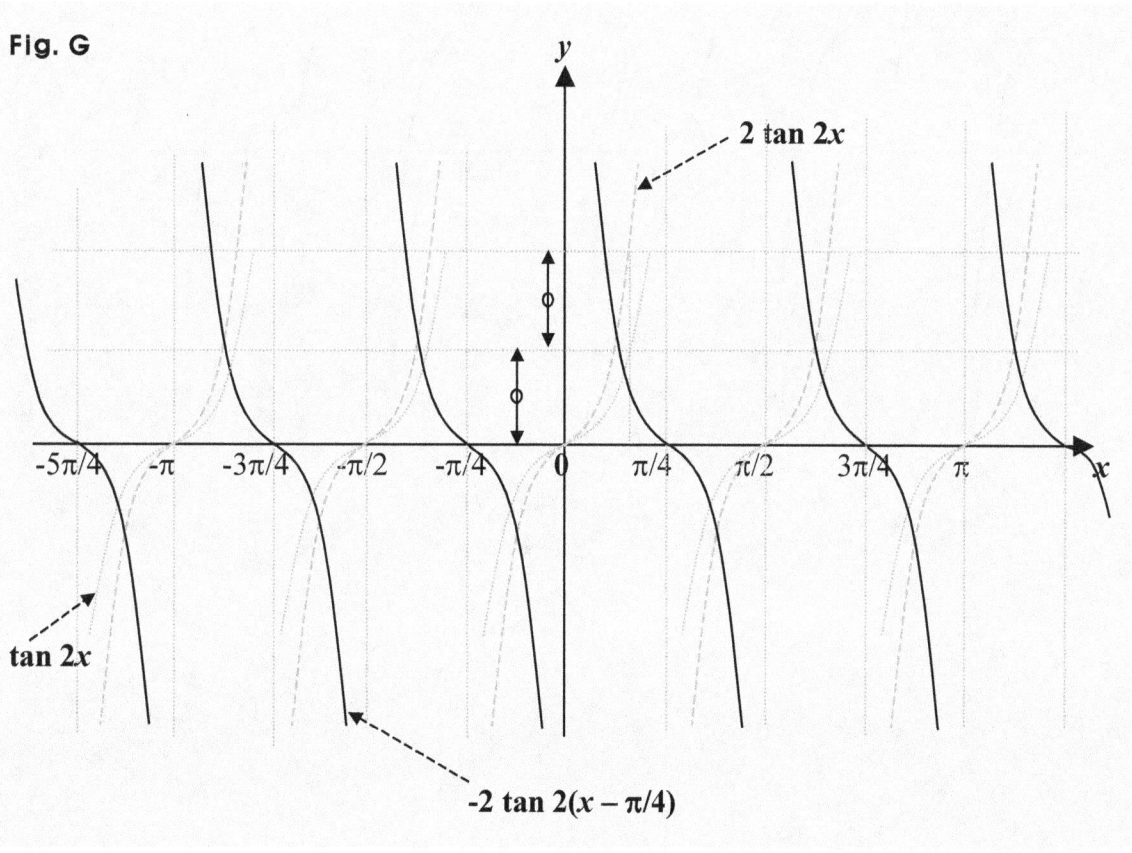

Fig. G

And next, setting the vertical shift to 1, we get: **-2tan (2x − π/4) + 1**.

Then, we move upward the curve with **-2tan 2(x − π/4)** in the amount of 1 to get the curve with **−2tan 2(x − π/4) + 1**, which is the curve of the function **T** given.

And thus, we can put in a graph the curve of the function **T** the way below:

74

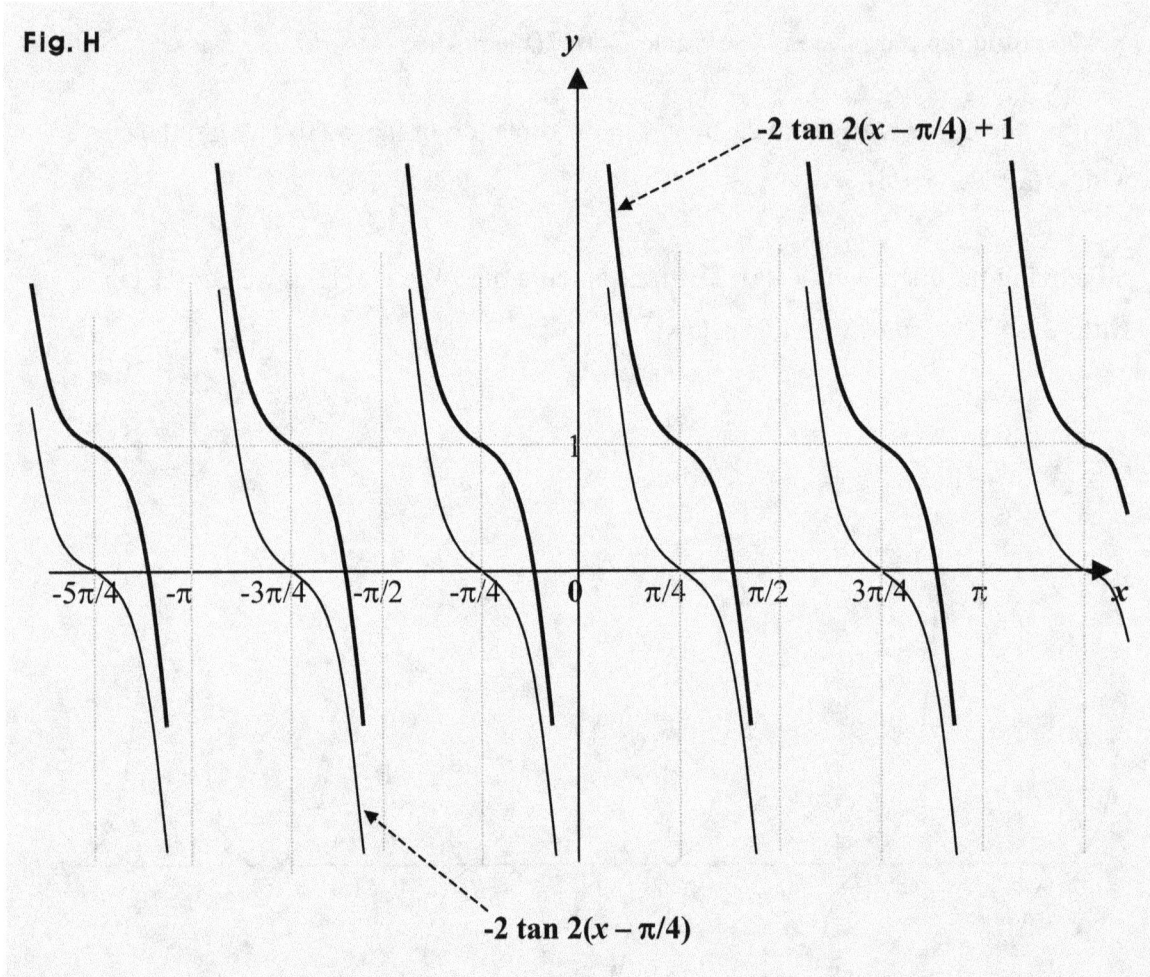

Fig. H

-2 tan 2(x − π/4) + 1

-2 tan 2(x − π/4)

Examples in Tangent Functions

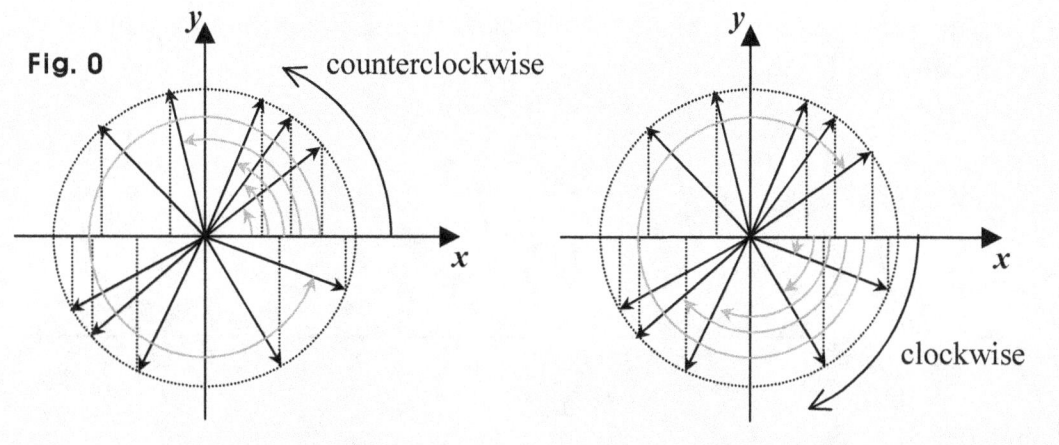

In trigonometry dynamic, the ray turning makes governing angles. If turning clockwise, it makes angles negative, and if counterclockwise, it makes angles positive.

And of course, if no turning, the angle made is 0. So a governing angle can be $0°$ or any angle positive or negative. In each right triangle, the ray is of length 1, and is the hypotenuse. And (x, y) is the terminal point, so x is the adjacent, and y is the opposite. And thus, assuming θ is a governing angle, we get: **$\tan \theta = y/x$**.

So if the ray is in the first quadrant, **$\tan \theta > 0$**, since both x and y are > 0.
In the second, **$\tan \theta < 0$**, since $x < 0$ but $y > 0$.
In the third, **$\tan \theta > 0$**, too, because both x and y are < 0.
And in the fourth, **$\tan \theta < 0$**, since $x > 0$ but $y < 0$.

Put in a graph the curve of each of the equations below:

0. $y = \frac{1}{2}\tan\left|\frac{x}{2}\right|$ 1. $|y| = \frac{1}{2}\tan\frac{x}{2}$ and $y = \left|\frac{1}{2}\tan\frac{x}{2}\right|$

2. $|y| = \frac{1}{2}\tan\left|\frac{x}{2}\right|$ 3. $y = \frac{1}{2}(\tan\frac{x}{2} + \tan\left|\frac{x}{2}\right|)$ 4. $|y| = \frac{1}{2}(\tan\frac{x}{2} + \tan\left|\frac{x}{2}\right|)$

Suggestions or Solutions

To the **Problem** in the Example **0**

Put in a graph the equation as follows: $y = \frac{1}{2}\tan|\frac{x}{2}|$.

To begin with, putting in the *x-y* plane, the curve of the tangent function $y = f(x) = \tan x$, we can put it the way below:

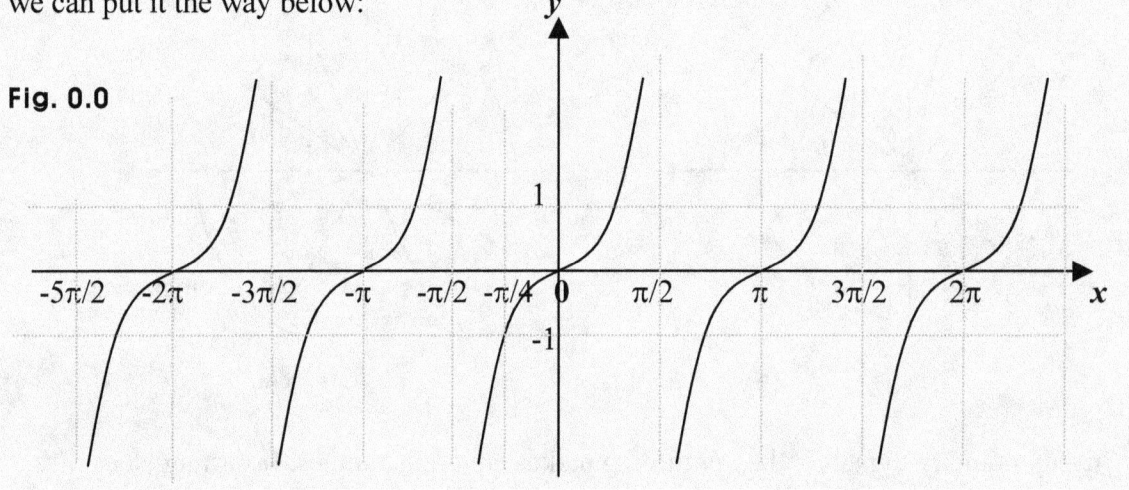

Fig. 0.0

The curve above is in fact, the same as the curve of an equation $y = \tan x$.

Notice that when $x = n\pi + \pi/4 = (4n + 1)\pi/4$, we get: $y = 1$. That is, $\tan(4n + 1)\pi/4 = 1$.

And the curve of an equation $y = \tan\frac{x}{2}$ is the curve of a function $y = g(x) = \tan\frac{x}{2}$, too.

So putting in a graph, the curve of the equation $y = \tan\frac{x}{2}$, we can put it this way:

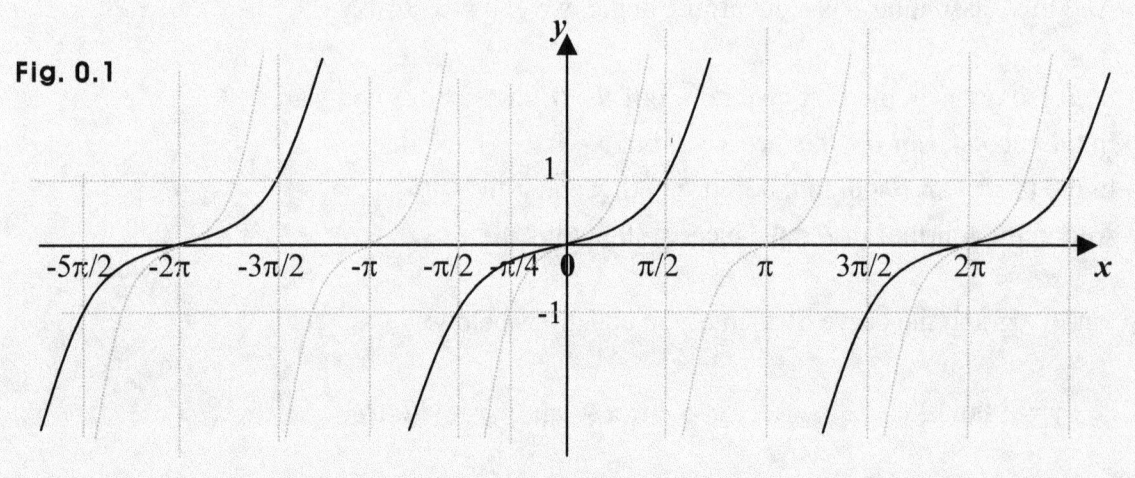

Fig. 0.1

And we can see that the period is 2π, and that the frequency is 1/2.

The frequency says how many times the periodic part appears in the basic period.
That is to say that the frequency is: the basic period over the particular period.

The basic period of the tangent, that is, the period of **tan** x is π.
And the particular period, that is, the period of **tan** $\frac{x}{2}$ is 2π.

So the frequency of the particular tangent, that is, the frequency of **tan** $\frac{x}{2}$ is: $\pi/2\pi = 1/2$.
And normally, setting $y = $ **tan** wx, we mean w is the frequency.

So looking at **tan** $\frac{x}{2}$, we can readily see that the frequency is $1/2$.
What then, about the curve of $y = \frac{1}{2}$ **tan** $\frac{x}{2}$?

For every value of x, the value of $\frac{1}{2}$ **tan** $\frac{x}{2}$ is half the value of **tan** $\frac{x}{2}$.
So we can put in a graph, the curve of $y = \frac{1}{2}$ **tan** $\frac{x}{2}$ the way below:

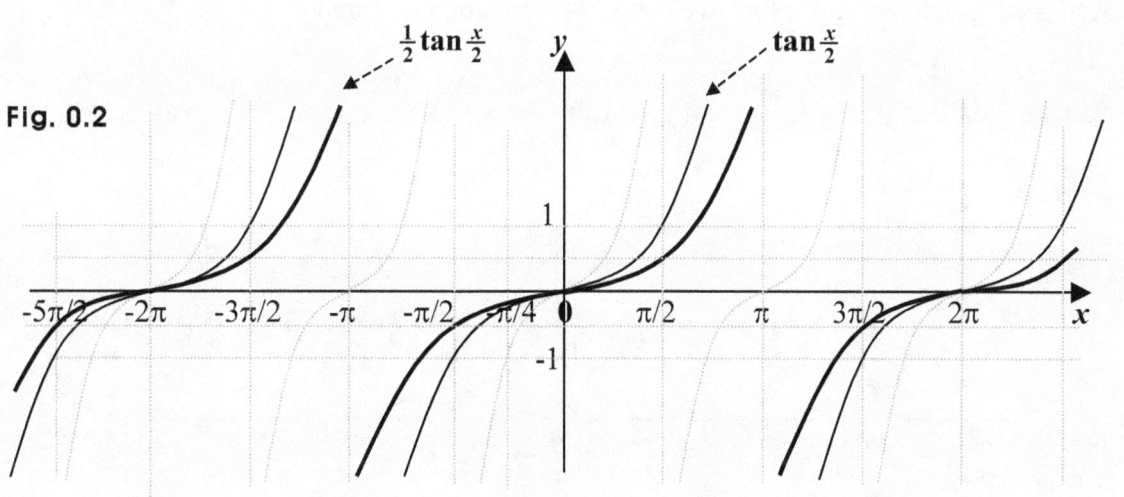

Fig. 0.2

What then, about $y = \frac{1}{2}$ **tan** $|\frac{x}{2}|$?

In that case, we want to consider two cases, one is: $x \geq 0$, and the other is: $x < 0$. Why?

We can put the equation this way, too: $y = \frac{1}{2}\tan\frac{1}{2}|x|$.

And since 1/2 is just a number, we may want to just consider this equation: $y = \tan|x|$.

And also, although looking awkward, **tan** $|x|$ can be put this way, too: **tan** t where $t = |x|$.

And we know: $|x| \geq 0$ for all x.

So even if $x < 0$, that is, x gets a negative value, we still get: $t > 0$, that is, t gets a positive value.

Thus, for instance, even if $x = -\pi/4$, we need to get: **tan** $\pi/4$, since $|-\pi/4| = \pi/4$. So given $y = \tan|x|$ for $x \geq 0$, we can just set: $y = \tan x$.

If however, $x < 0$, we want to set: $y = \tan(-x)$ because $-x > 0$ since $|x| \geq 0$ in **tan** $|x|$.

And we have a trig-identity: **tan** $(-x) = $ -**tan** x.

And thus, given $y = \frac{1}{2}\tan\frac{1}{2}|x|$, we get: $y = \frac{1}{2}\tan\frac{x}{2}$ for $x \geq 0$, and $y = -\frac{1}{2}\tan\frac{x}{2}$ for $x < 0$.

So taking care of first, the curve of $y = \frac{1}{2}\tan\frac{x}{2}$ for $x \geq 0$, we can put it the way below:

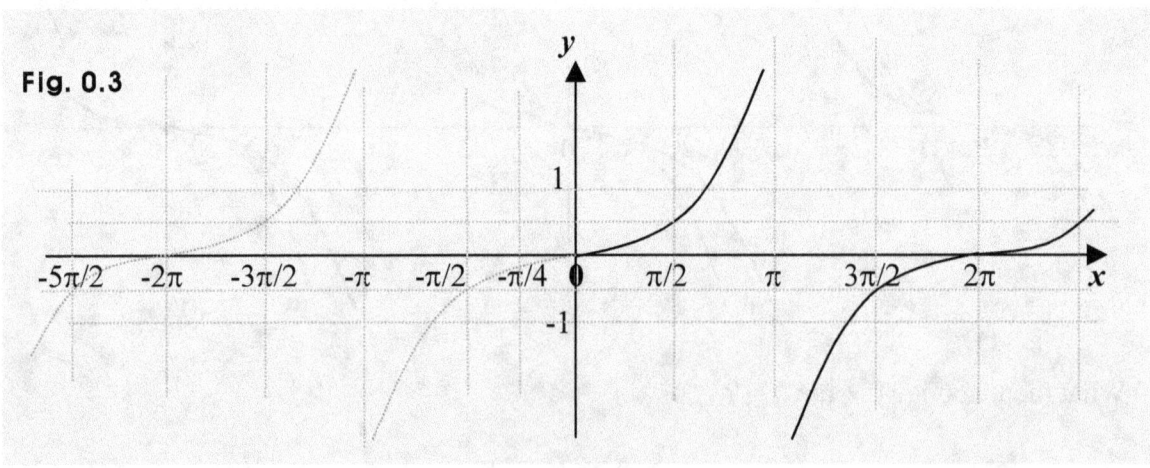

Fig. 0.3

Next, putting in a graph, the curve of $y = -\frac{1}{2}\tan\frac{1}{2}|x|$, we can put it the way below:

Fig. 0.4

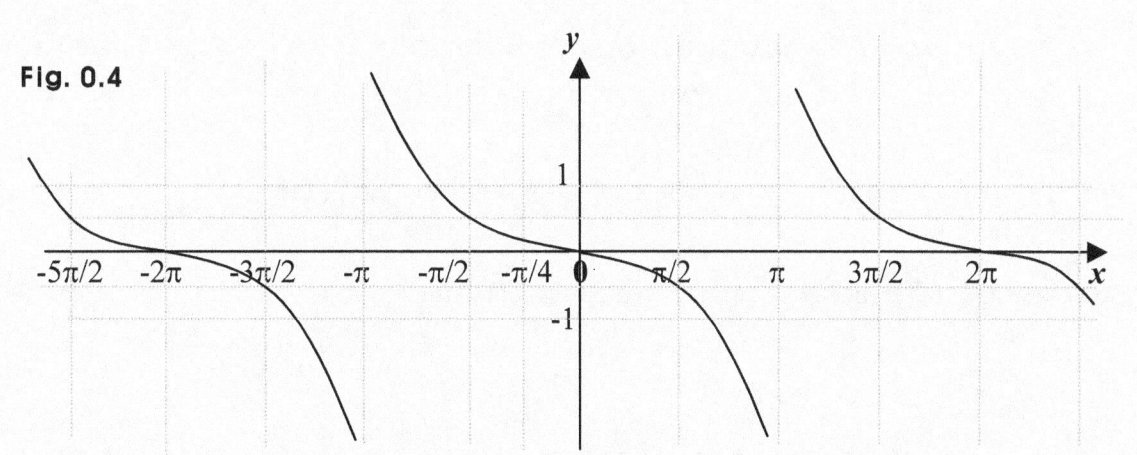

So moving on to the curve of $y = -\frac{1}{2}\tan\frac{x}{2}$ for $x < 0$, we can put it the way below:

Fig. 0.5

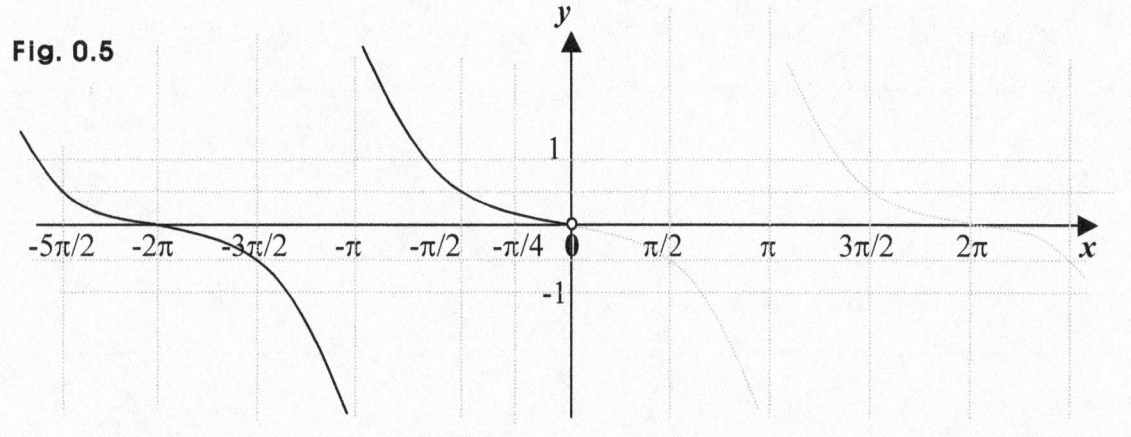

And thus, putting in a graph, the curve of $y = \frac{1}{2}\tan\left|\frac{x}{2}\right|$, we can put it the way below:

Fig. 0.6

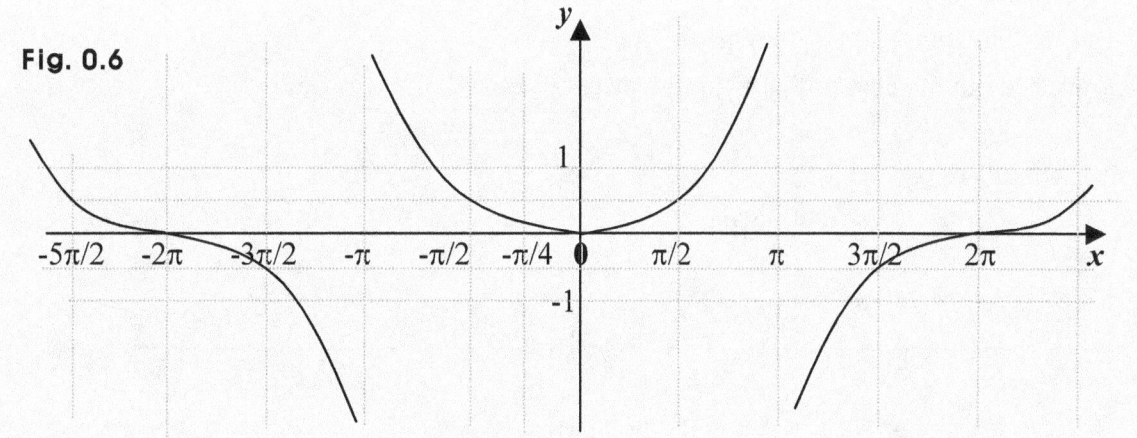

Notice that the curve is symmetric about the y-axis.

In fact, the curve of an equation or a function $y = f(|x|)$ is symmetric about the y-axis.

Suggestions or Solutions
To the **Problem** in the Example **1**

Put in a graph the equations as follows: $|y| = \frac{1}{2}\tan\frac{x}{2}$ and $y = |\frac{1}{2}\tan\frac{x}{2}|$.

So what's the difference between the two equations?

Other than the looks, they are the same. So both are the same.

Thus for instance, an equation $y = |\tan x|$ is the same as an equation $|y| = \tan x$.

And we know $|y| \geq 0$. So we get: $y = \frac{1}{2}\tan\frac{x}{2}$ for $y \geq 0$, and $y = -\frac{1}{2}\tan\frac{x}{2}$ for $y < 0$.

And putting in a graph, the curve of $y = \frac{1}{2}\tan\frac{x}{2}$ for $y \geq 0$, we can put it the way below:

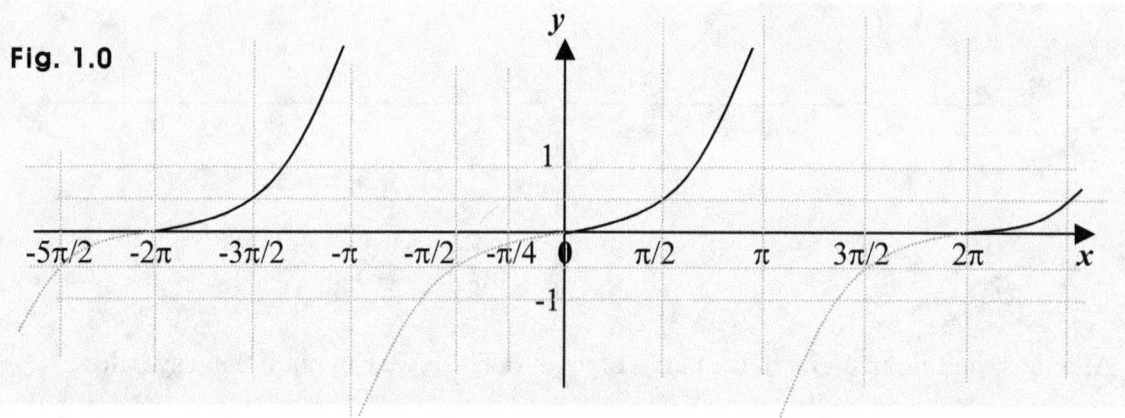

Fig. 1.0

And we know the curve symmetric about the y-axis.

So we can put the curve of $|y| = \frac{1}{2}\tan\frac{x}{2}$ or $y = |\frac{1}{2}\tan\frac{x}{2}|$, the way below:

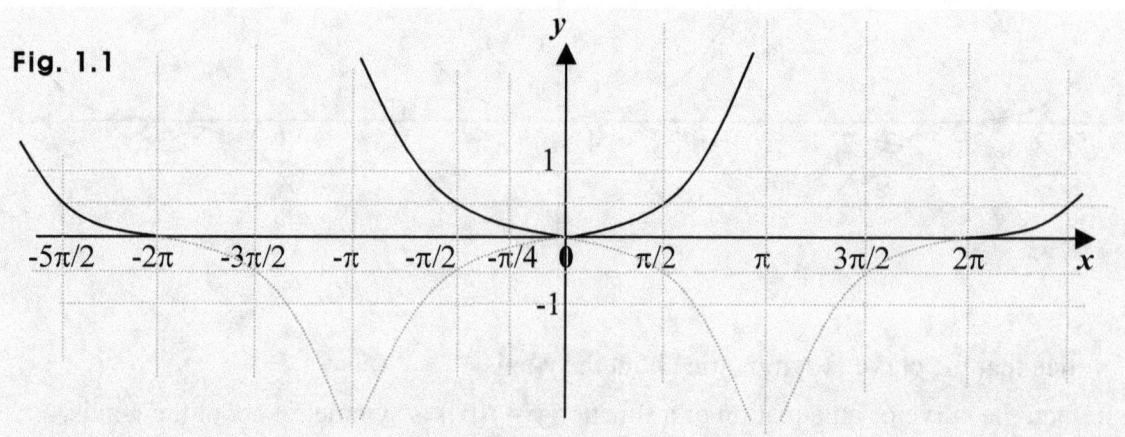

Fig. 1.1

Suggestions or Solutions
To the **Problem** in the Example **2**

Put in a graph the equations as follows: $|y| = \frac{1}{2}\tan|\frac{x}{2}|$.

To begin with, putting in the *x-y* plane, the curve of the equation $y = \frac{1}{2}\tan\frac{x}{2}$, we can put it the way below:

Fig. 2.0

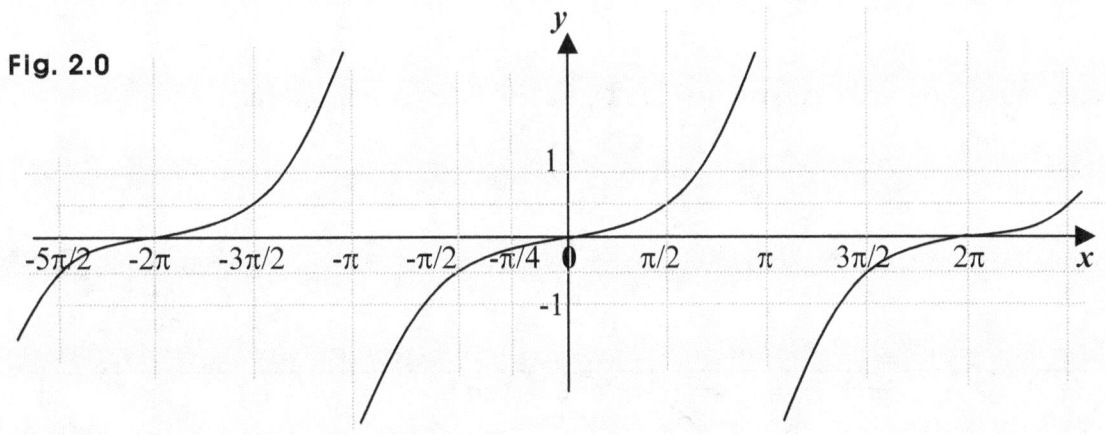

Next, we know: $|y| \geq 0$ for all *y*, and $|2x| \geq 0$ for all *x*.
So we want to consider four different cases as follows:

First, assuming $x \geq 0$, and $y \geq 0$, we get: $y = \frac{1}{2}\tan\frac{x}{2}$.

Next, assuming $x \geq 0$, and $y < 0$, we get: $-y = \frac{1}{2}\tan\frac{x}{2}$, so we get: $y = -\frac{1}{2}\tan\frac{x}{2}$.

Next, assuming $x < 0$, and $y < 0$, we get: $-y = \frac{1}{2}\tan(-\frac{x}{2}) = -\frac{1}{2}\tan\frac{x}{2}$, so we get: $y = \frac{1}{2}\tan\frac{x}{2}$.

And next, assuming $x < 0$, and $y \geq 0$, we get: $y = \frac{1}{2}\tan(-\frac{x}{2})$, so we get: $y = -\frac{1}{2}\tan\frac{x}{2}$.

Thus in sum, we have:

$y = \frac{1}{2}\tan\frac{x}{2}$ for $x \geq 0$, and $y \geq 0$. $\qquad$ $y = \frac{1}{2}\tan\frac{x}{2}$ for $x < 0$, and $y < 0$.

$y = -\frac{1}{2}\tan\frac{x}{2}$ for $x \geq 0$, and $y < 0$. $\qquad$ $y = -\frac{1}{2}\tan\frac{x}{2}$ for $x < 0$, and $y \geq 0$.

So beginning with the curve of $y = \frac{1}{2}\tan\frac{x}{2}$ for $x \geq 0$, and $y \geq 0$, we get:

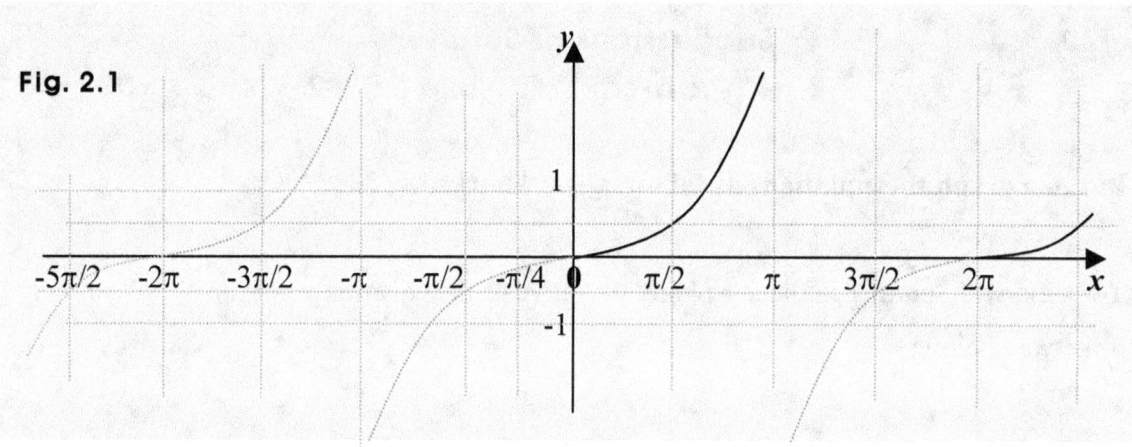

Fig. 2.1

Next, moving on to the curve of $y = \frac{1}{2}\tan\frac{x}{2}$ for $x < 0$, and $y \geq 0$, we get:

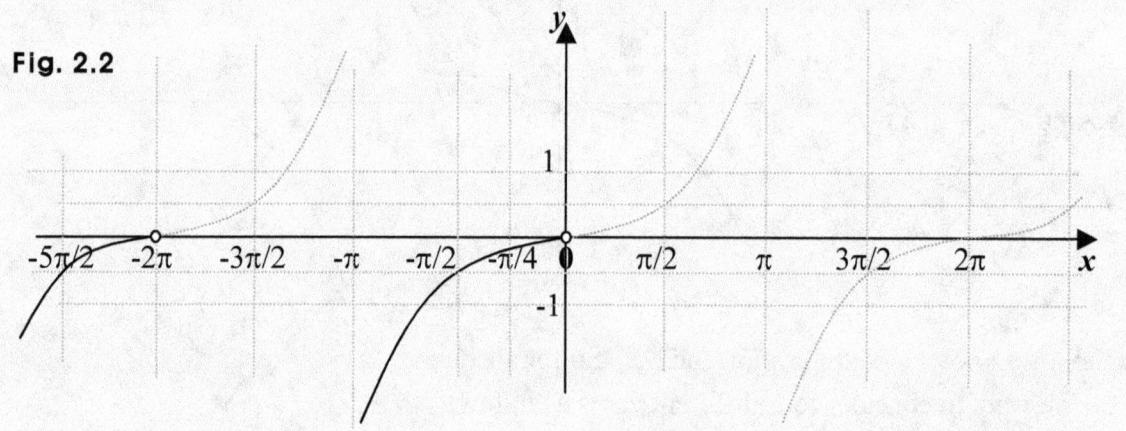

Fig. 2.2

Next, we can put in a graph, the curve of $y = -\frac{1}{2}\tan\frac{x}{2}$ the way below:

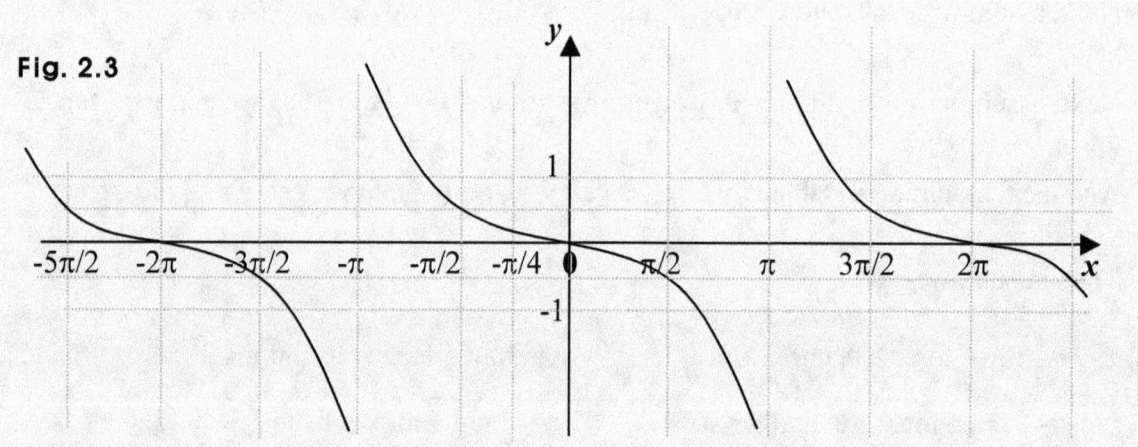

Fig. 2.3

So next, moving on to the curve of $y = -\frac{1}{2}\tan\frac{x}{2}$ for $x \geq 0$, and $y < 0$, we get:

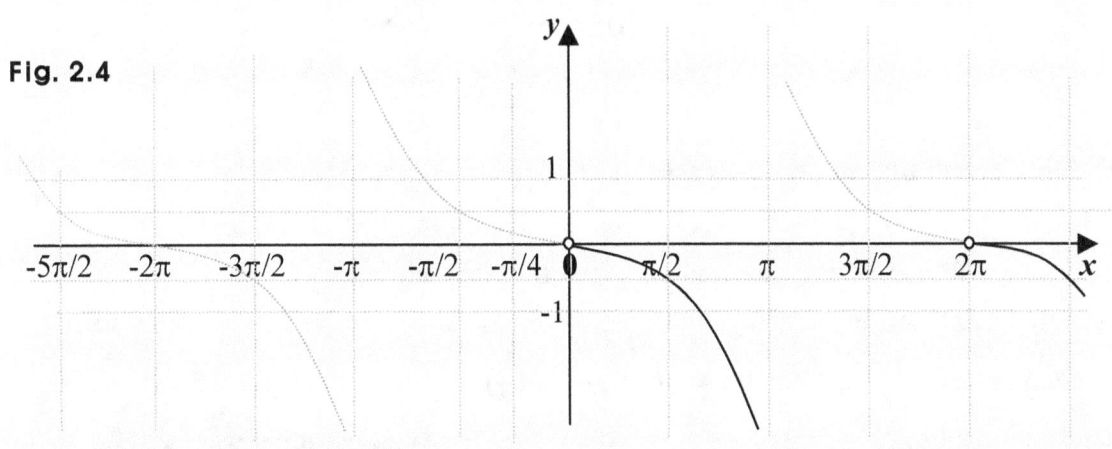

Fig. 2.4

And next, moving on to the curve of $y = -\frac{1}{2}\tan\frac{x}{2}$ for $x < 0$, and $y \geq 0$, we get:

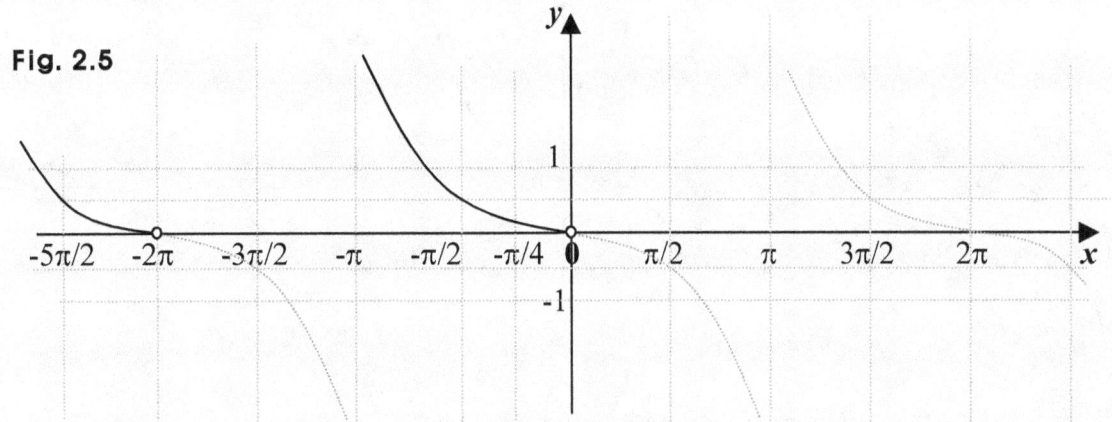

Fig. 2.5

And thus, putting in a graph, the curve of $|y| = \frac{1}{2}\tan\left|\frac{x}{2}\right|$, we can put it the way below:

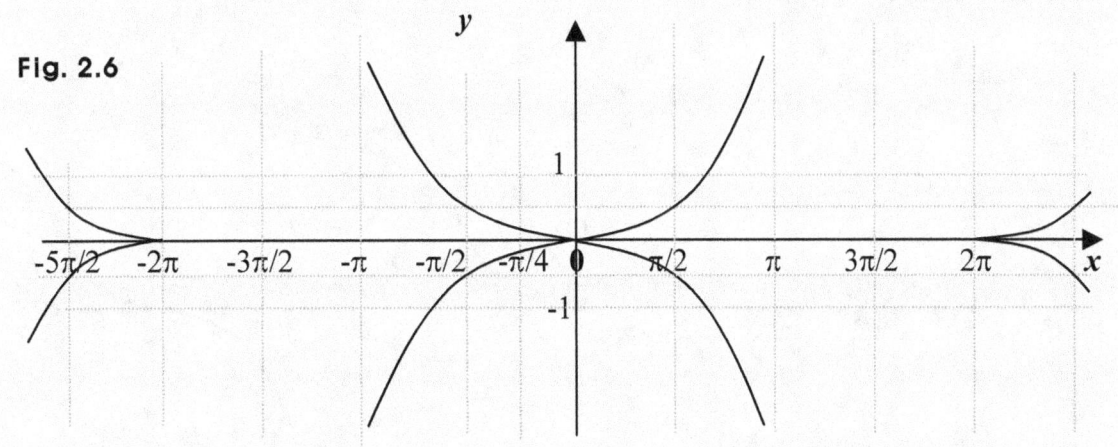

Fig. 2.6

Notice that the curve is symmetric about the origin as well as the x and y axes.

Suggestions or Solutions
To the **Problem** in the Example **3**

Put in a graph the equations as follows: $y = \frac{1}{2}(\tan\frac{x}{2} + \tan|\frac{x}{2}|)$.

To begin with, we can put the equation given the way below:

Assuming first, $x \geq 0$, we get: $y = \frac{1}{2}(\tan\frac{x}{2} + \tan\frac{x}{2}) = \tan\frac{x}{2}$. So we get: $y = \tan\frac{x}{2}$ for $x \geq 0$.

Assuming next, $x \backslash < 0$, we get: $y = \frac{1}{2}(\tan\frac{x}{2} - \tan\frac{x}{2}) = 0$. So we get: $y = 0$ for $x < 0$.

Thus next, putting in a graph, the curve of $y = \tan\frac{x}{2}$, we can put it the way below:

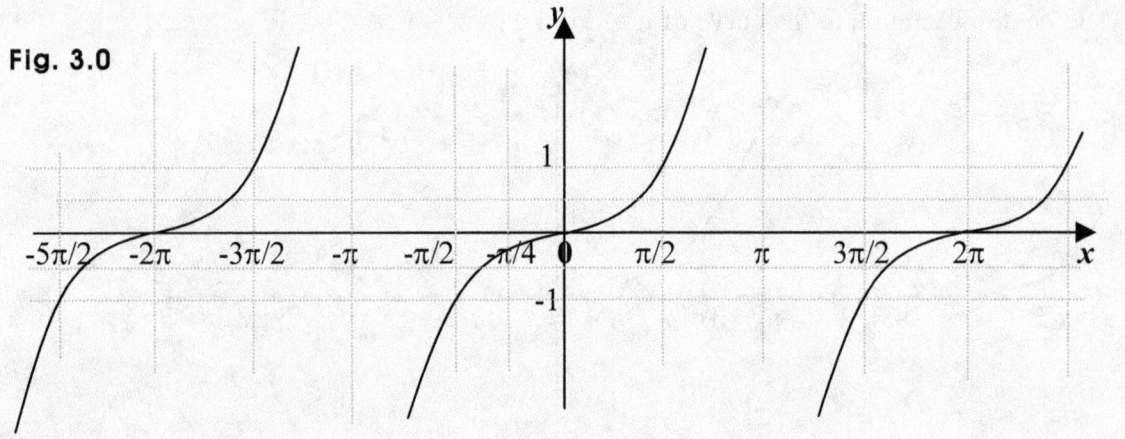

Fig. 3.0

Notice that when $x = 2n\pi + \pi/2$ for n integer, we get: $y = 1$.

That is, we get: $\tan(4n + 1)\pi/2 = 1$ for n integer.

And thus, putting in a graph, the curve of $y = \frac{1}{2}(\tan\frac{x}{2} + \tan|\frac{x}{2}|)$, we can put it the way

below:

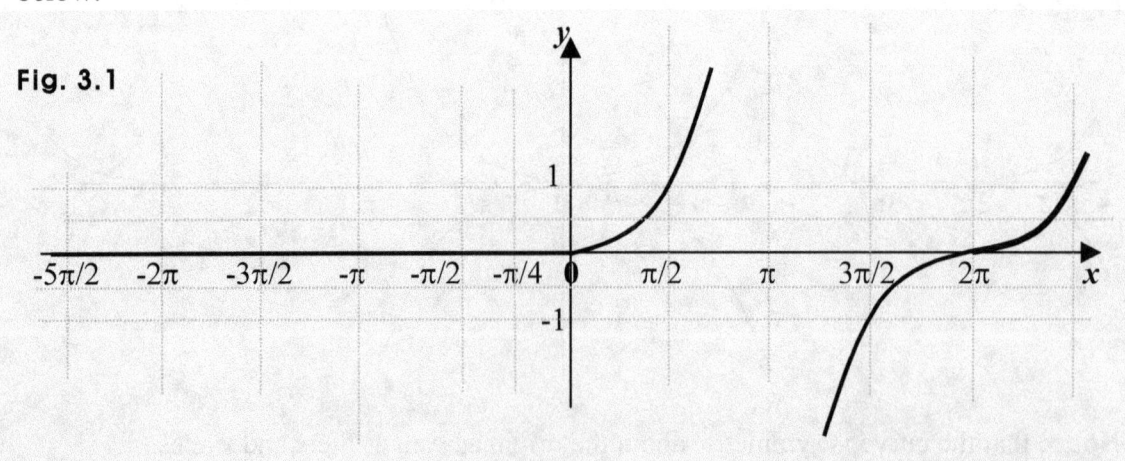

Fig. 3.1

And of course, we can get the same using graphs. So to begin with, putting in a graph, the curve of $y = \frac{1}{2}\tan\frac{x}{2}$, we can put it the way below:

Fig. 3.2

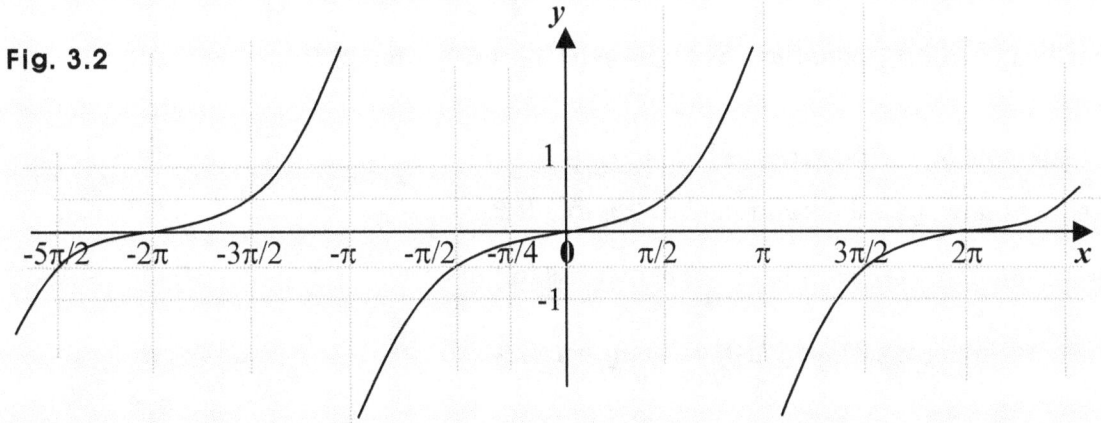

Next, putting in a graph, the curve of $y = \frac{1}{2}\tan\left|\frac{x}{2}\right|$, we can put it the way below:

Fig. 3.3

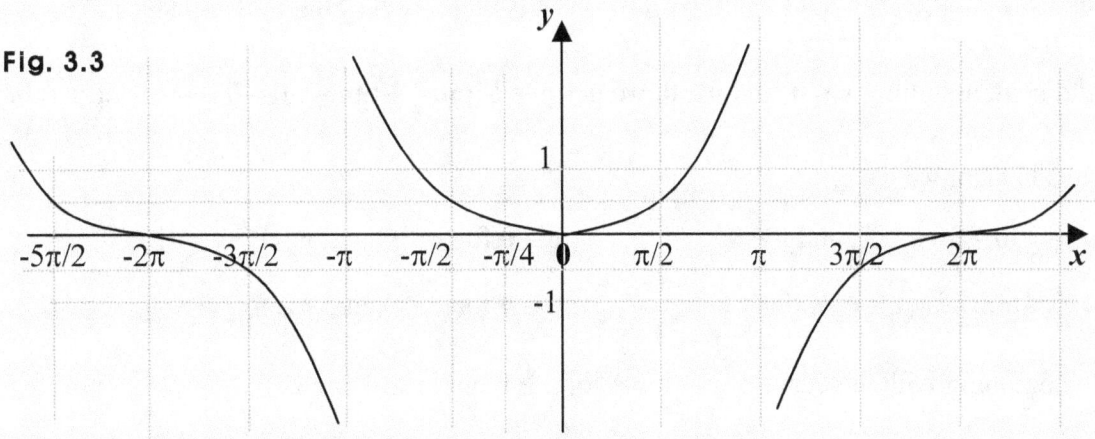

And thus, adding together the two curves above, we get:

Fig. 3.4

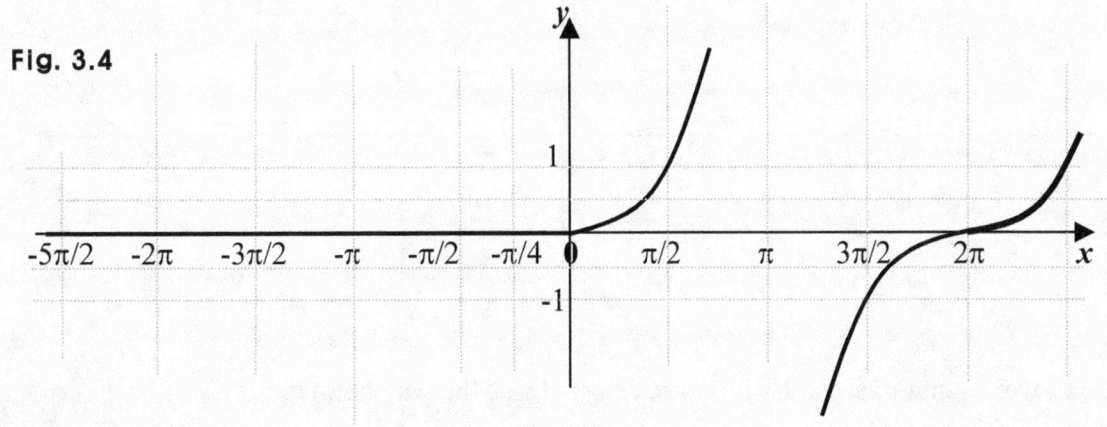

Suggestions or Solutions
To the Problem in the Example 4

Put in a graph the equations as follows: $|y| = \frac{1}{2}(\tan\frac{x}{2} + \tan|\frac{x}{2}|)$.

To begin with, we know: $\tan|\frac{x}{2}| = \tan\frac{1}{2}|x|$.

And we know: $|y| \geq 0$ for all y, and $|x| \geq 0$ for all x.

So we want to consider four different cases as follows:

First, assuming $x \geq 0$, and $y \geq 0$, we get: $y = \tan\frac{x}{2}$.

Next, assuming $x \geq 0$, and $y < 0$, we get: $-y = \tan\frac{x}{2}$, so we get: $y = -\tan\frac{x}{2}$.

Next, assuming $x < 0$, and $y < 0$, we get: $-y = \frac{1}{2}\{\tan\frac{x}{2} + \tan(-\frac{x}{2})\} = 0$, so we get: $y = 0$.

And next, assuming $x < 0$, and $y \geq 0$, we get: $y = \frac{1}{2}\{\tan\frac{x}{2} + \tan(-\frac{x}{2})\} = 0$, so we get: $y = 0$.

Thus in sum, we have:

$y = \tan\frac{x}{2}$ for $x \geq 0$, and $y \geq 0$. $\qquad y = 0$ for $x < 0$, and $y < 0$.

$y = -\tan\frac{x}{2}$ for $x \geq 0$, and $y < 0$. $\qquad y = 0$ for $x < 0$, and $y \geq 0$.

So beginning with the curve of $y = \tan\frac{x}{2}$ for $x \geq 0$, and $y \geq 0$, we get:

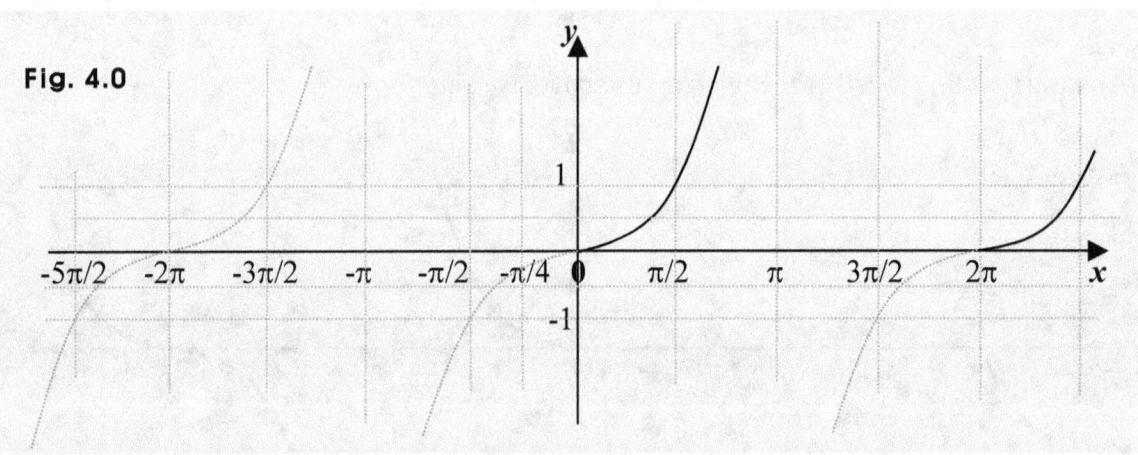

Fig. 4.0

Next, we can put in a graph, the curve of $y = -\tan\frac{x}{2}$ the way below:

Fig. 4.1

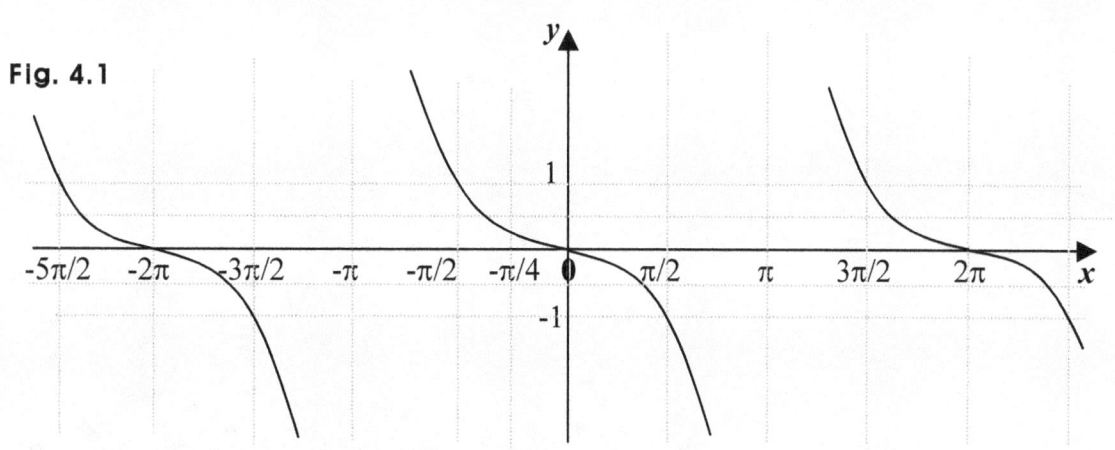

So next, moving on to the curve of $y = -\tan\frac{x}{2}$ for $x \geq 0$, and $y < 0$, we get:

Fig. 4.2

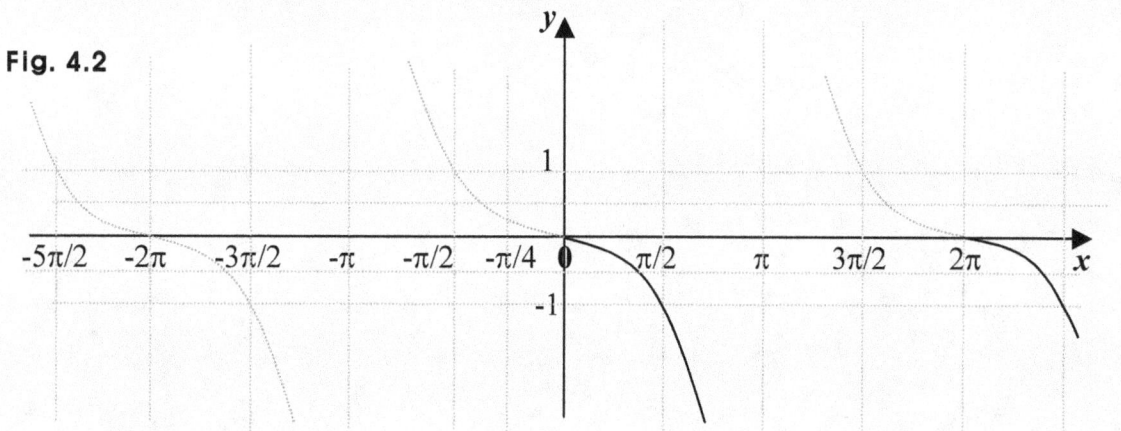

And we have: $y = 0$ for $x < 0$, and $y < 0$, and also, $y = 0$ for $x < 0$, and $y \geq 0$.

And thus, we can put in a graph, the curve of $|y| = \frac{1}{2}(\tan\frac{x}{2} + \tan|\frac{x}{2}|)$ the way below:

Fig. 4.3

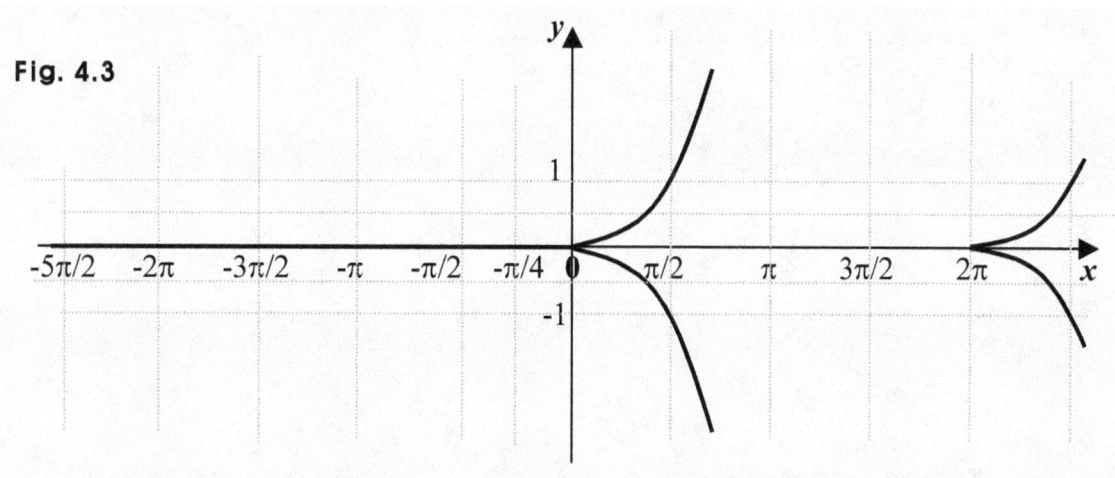

D. The Reciprocals

What reciprocals?

They are the reciprocals of the basic trig-ratios as the sine.

So for instance, taking the reciprocal of the sine, we get the cosecant, denoted by **csc**. Taking thus, the reciprocal of the sine, that is, **sin x**, we get: **csc x**.

And we can take **sin x** as a function, can set: $y = f(x) = $ **sin x**, and call it a sine function. What then, about a *cosecant function*?

We can make another trig-function called a cosecant function using **csc x**, of course. So for instance, we can make cosecant functions as $y = g(x) = $ **csc x** or $y = h(x) = $ **2csc x**.

However, the cosecant functions above cannot be defined for all angles. How come?

We know that <u>the sine is: the opposite over the hypotenuse</u>.

So <u>the cosecant is: the hypotenuse over the opposite</u>, since it is the reciprocal of the sine.

And we know that the opposite can be 0 for some angles as 0, -π, and π.

So for instance, the function **csc x** cannot be defined for $x = \pi$. And in general, it cannot be defined for $x = n\pi$ where n is an integer.

So just setting: $y = g(x) = \csc x$, we mean that $x \neq n\pi$ where n is an integer.

That is, the domain is a set of all real numbers less $n\pi$ for n integer.

So we can define or specify the cosecant function the way below, too:

$y = g(x) = \csc x$ for $x \neq n\pi$ for n integer.

What then, about its curve?

We know the cosecant is the reciprocal of the sine. So keeping the fact in mind, and using the curve of the sine function $y = f(x) = \sin x$, we can put in a graph, the curve of the cosecant function g the way below:

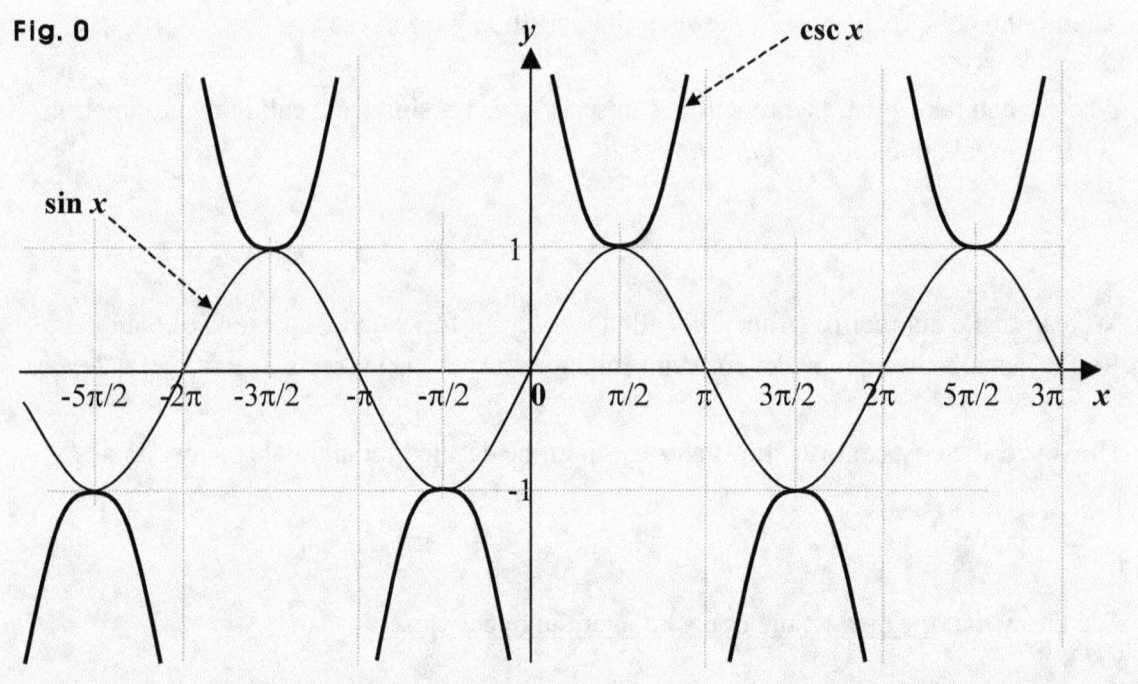

So we can now see that the y-values, that is, the outputs are grater than or equal to 1 or less than or equal to –1. And thus, the range is: $y \leq \textbf{-1}$ or $y \geq \textbf{1}$, that is, $|y| \geq \textbf{1}$.

And of course, if the domain is different, the range can be other than the one above. Assuming for instance, $y = h(x) = \csc x$ for $0 < x < \pi$, we get: $y \geq \textbf{1}$, which is the range.

And the cosecant function $y = g(x) = \csc x$ can be called the prototype, and is thus, in the most basic form. Assuming G is a cosecant function, too, and the domain is a set of all real numbers less $n\pi$ for n integer, and using a general form, we can put it the way as follows:

$y = G(x) = A \cdot \csc \{w(x + a)\} + b$ where $w(x + a) \neq n\pi$ for n integer, and A, w, a, and b are constant.

And just simply, we often put it this way, too: $y = G(x) = A \cdot \csc w(x + a) + b$.

Then, $|w|$ is called the frequency, $\frac{2\pi}{|w|}$ is the period, and a is called the phase.

So in the function $y = g(x) = \csc x$, the phase is 0, and the frequency is 1, so the period is 2π. What do we mean by though, the period?

Looking at the curve above, we can see that a part of the cuve repleats itself. The part is in two pieces though.
And the part is in fact, the smallest part repeating itself. What then, is the part?

It is the part from 0 to 2π. So the part repeats itself every 2π interval. And we call such an interval a period.
So the interval 2π is the period in the cosecant function $y = g(x) = \csc x$.

• And thus, we call a cosecant function a *periodic* function, too.

And the cosecant function g above is the prototype, and thus, is in the most basic form.

• So the period 2π can be called the *basic period* in *cosecant functions*. In other words:

• The interval 2π can be called the *basic interval* in *cosecant functions*.

And for instance, in the general form, setting w to 2, a and b to 0 each, and A to 1, we get a new function where: $y = k(x) = \csc 2x$.
What then, is its domain?

We know that **csc θ** is not defined for $\theta = n\pi$ where n is an integer.

So in **csc 2x**, we don't want **2x** to be **$n\pi$**.

That is, we want: **$2x \neq n\pi \Rightarrow x \neq n\pi/2$**, which means x is not an integer multiple of 90°.

And thus, the domain of the function **$k(x) = $ csc 2x** can be a set of all real numbers less **$n\pi/2$** for **n** integer. And of course, if **E** is the domain just described, **E** is the biggest domain that the cosecant function **csc 2x** can have.

So putting in the **x-y** plane, the curve of the cosecant function $y = k(x) = $ **csc 2x**, we get:

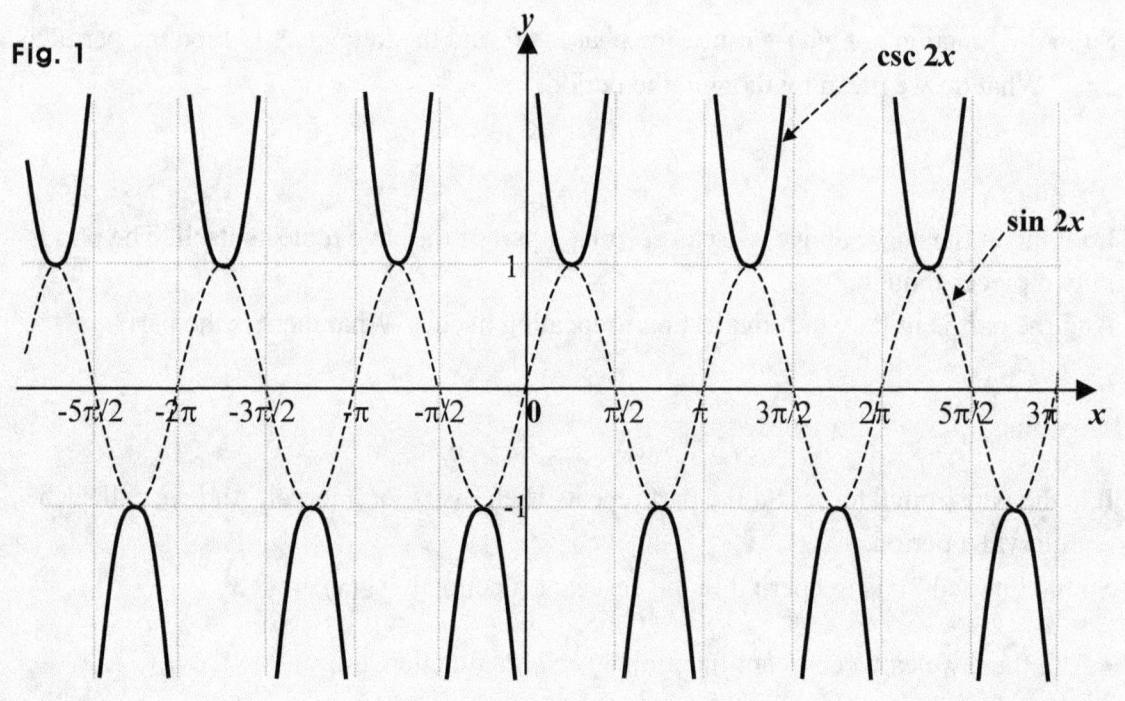

Fig. 1

(Note that not just one but all the curved lines in solid black form the curve of **csc 2x**.)

Then again, we can see that a part of the cuve repleats itself, and is in two pieces. And the part is of course, the smallest part repeating itself.   What then, is the part?

It is the part from 0 to π. So the part repeats itself every π interval.

And we call such an interval a period. So the interval π is the period in the function **k**.

And the smallest part stated above appears twice in the basic interval 2π, which is the basic period.

Then, we say that the frequency is 2 in the function k.

So the frequency is the number of times the smallest part repeating itself appears in the basic inverval 2π, which is the basic period.

• In short, the frequency is the number of times the smallest part appears in 2π.

More precisely though, we want to put the frequency the way below:

   • In case of cosecant functions, the frequency is the number of times the samllest part repeating itself appears in the basic period or the basic interval 2π.

And thus, in the general form $A \cdot \csc w(x + a) + b$, $|w|$ is the frequency.
Then, given a cosecant function in the general form, how can we get its period?

We know that $|w|$ is the frequency, that is, the number of times the smallest part repeating itself appears in the basic interval (period) 2π, and that the interval that fits the smallest part is the period.

So using the frequency $|w|$ and the basic interval 2π, we can put the period this way: $\frac{2\pi}{|w|}$.

And we can have a cosecant function where $y = h(x) = \csc(-x)$ for $x \neq n\pi/2$ for n integer.

Then, the frequency in the cosecant function h is $|-1|$, simply because a frequency is positive.

And we can put h this way, too: $y = h(x) = -\csc x$ for $x \neq n\pi/2$ for n integer.

It's because $\csc(-x) = -\csc x$.

And we know the cosecant is: the hypotenuse over the opposite.
So assuming the ray, that is, the hypotenuse is of length 1, we get:

Fig. 2

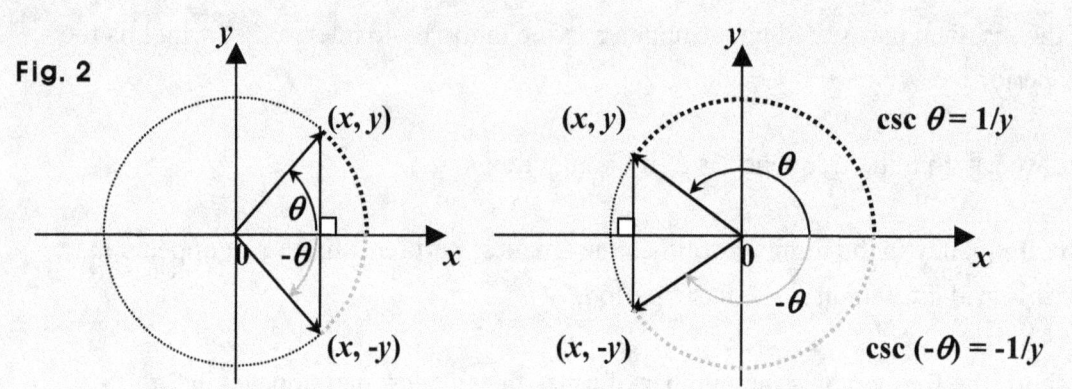

So we get: **csc (-θ) = -csc θ.**   And putting in the **x-y** plane, the curve of **-csc x**, we get:

Fig. 3

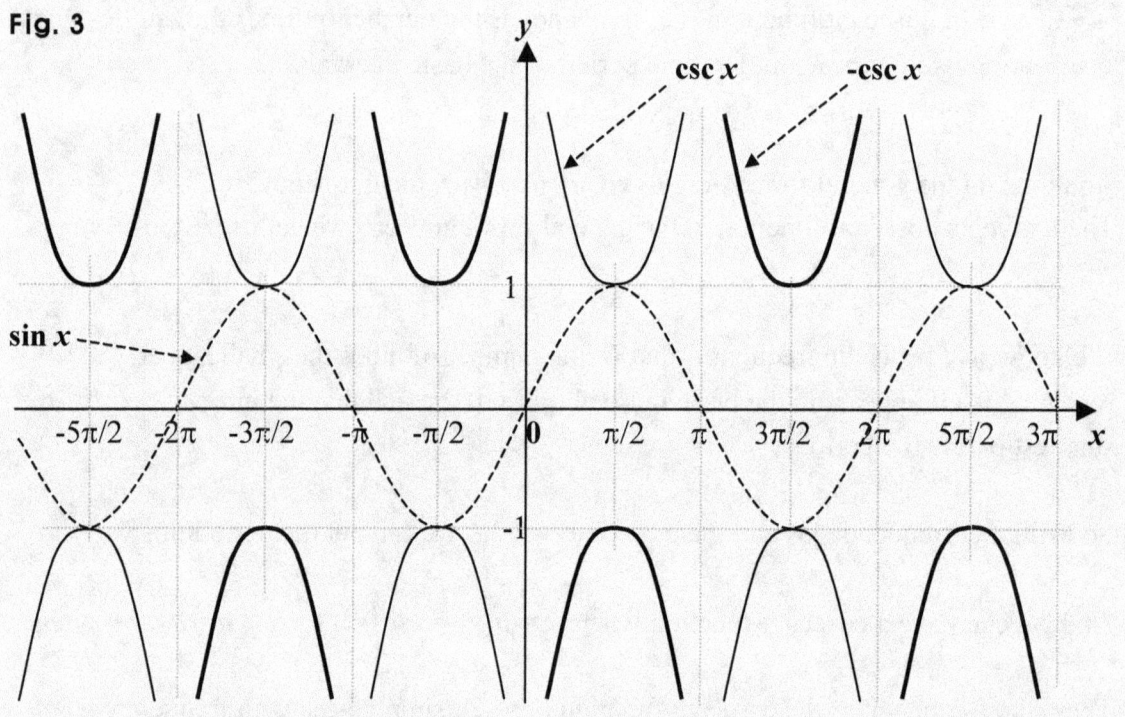

And we can see every 2π interval, the smallest part repeating itself, which is in two pieces though. So the period of $\underline{h(x) = -csc\ x}$ is 2π, too.

• And next, what do we mean by the phase?

Using the general form, we can put a cosecant function called **G** the way below:

$y = G(x) = A \cdot csc\ w(x + a) + b$, where $w(x + a) \neq n\pi$ for **n** integer.

Then, a is called the phase, $|w|$ is called the frequency, and $\frac{2\pi}{|w|}$ is the period.

Now, shifting by $-a$ in the direction of the x-axis, the curve of $y = Q(x) = A \cdot \csc wx + b$ where $wx \neq n\pi$ for n integer, we get the curve of the function G above.

So the two curves themselves of G and Q are the same, and moving the curve of Q in the amount of $-a$ along the x-axis, we get the curve of G. And if the phase is positive, the curve gets shifted to the left, and if negative, it gets shifted to the right.

Assuming for instance, shifting the curve of $p(x) = \csc x$ for $-\pi < x < 3\pi$ by $-\pi/2$ in the direction of the x-axis, that is, to the left, we get the curve of a function below:

$q(x) = \csc (x + \pi/2)$ for $-3\pi/2 < x < 5\pi/2$. So putting the two curves in one graph, we get:

Fig. 4

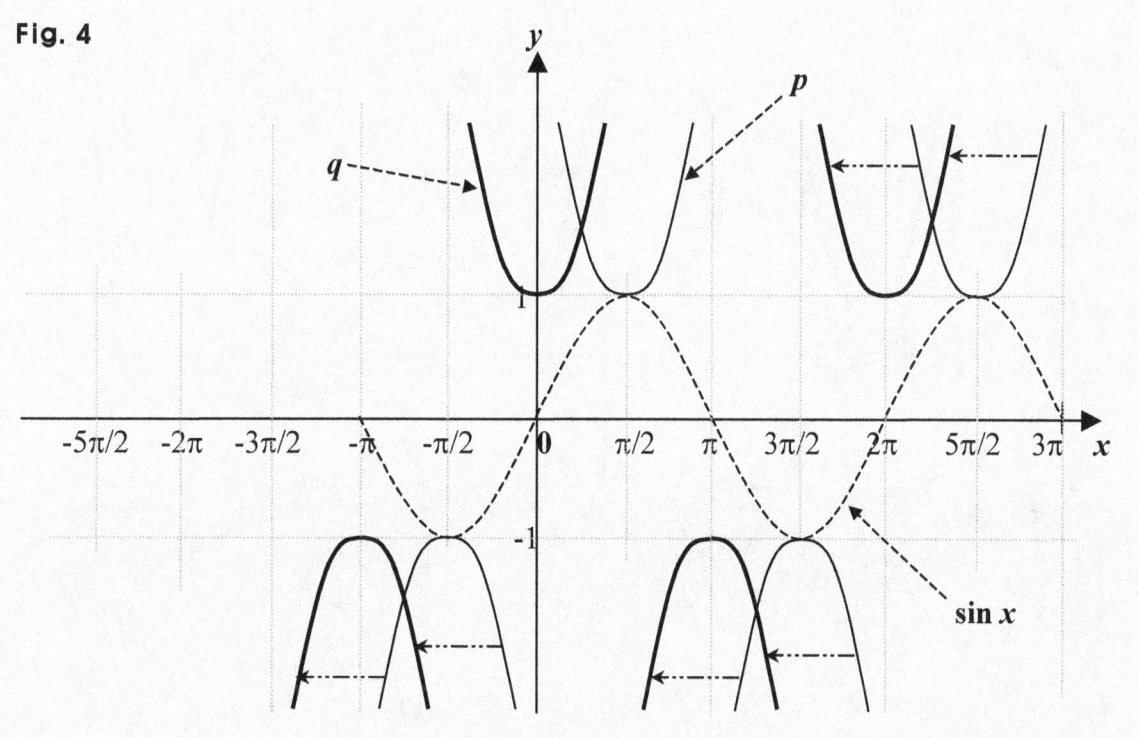

And thus, the curve of p gets moved to the left in the amount of $\pi/2$. And the new curve is the curve of q. So we can call the phase a horizontal shift, too.

What then, about the constant b in $y = Q(x) = A \cdot \csc wx + b$?

It is called a vertical shift.

So shifting the curve of a function $y = V(x) = A \cdot \csc w(x + a)$ by b in the direction of the y-axis, we get the curve of the function G, which is: $y = G(x) = A \cdot \csc w(x + a) + b$.

So the two curves themselves of G and V are the same, and moving the curve of V in the amount of b along the y-axis, we get the curve of G.

And if b is positive, the curve is shifted upward, and if negative, it moves downward.

Assuming for instance, shifting by 1/2 in the direction of the y-axis, that is, upward, the curve of $u(x) = \csc x$, we get the curve of a function as follows: $v(x) = \csc x + 1/2$.

Putting therefore, the two curves in a graph, we get:

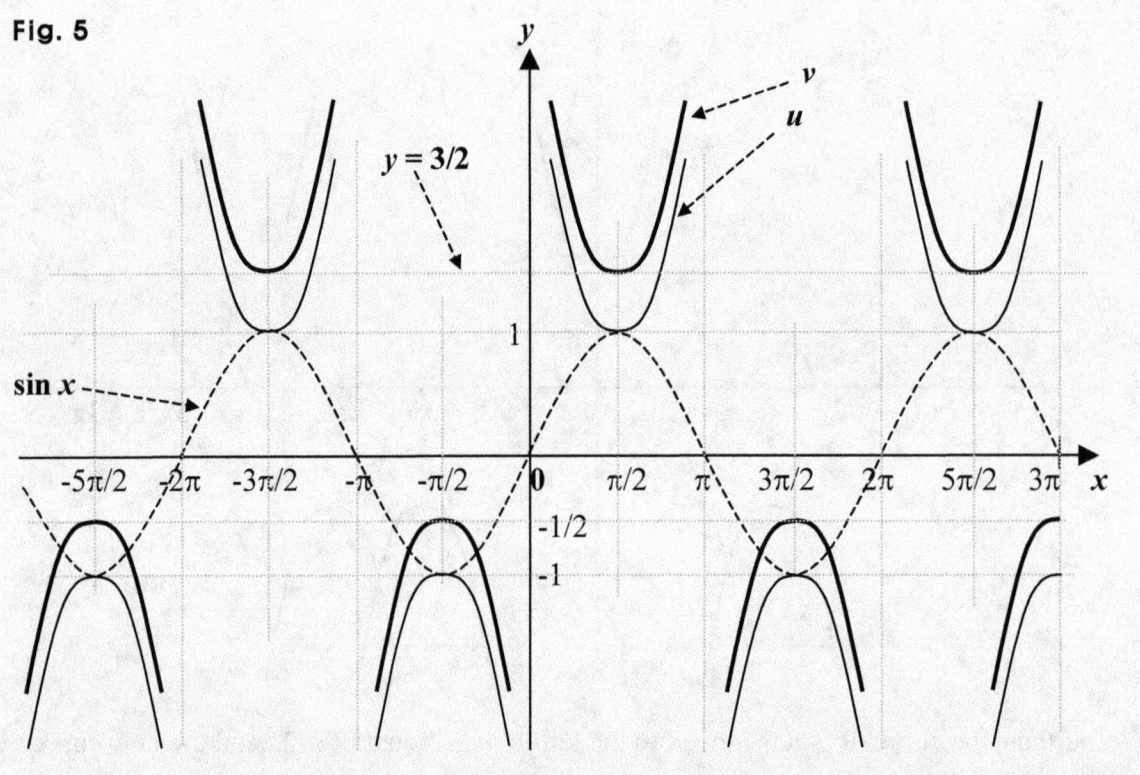

Fig. 5

So the curve of u gets moved upward in the amount of 1/2. And the new curve is the curve of v. So we can call the constant b a vertical shift.

And let's now put in a graph, for instance, the curve of a cosecant function below:

$y = S(x) = (-1/2)\csc(-2x + \pi) + 1$, where $-2x + \pi \neq n\pi$ for n integer.

To begin with, we have: $\csc(-\theta) = -\csc\theta$.

So we can get: $\csc(-2x + \pi) = \csc\{-(2x - \pi)\} = -\csc(2x - \pi)$.

Thus, we get: $(-1/2)\csc(-2x + \pi) + 1 = (1/2)\csc(2x - \pi) + 1$.

And putting it in the general form, we get: $(1/2)\csc 2(x - \pi/2) + 1$.

So we can now see that the frequency is 2, the phase is $-\pi/2$, that is, shifting by $\pi/2$ to the right, and the vertical shift is 1.

And next, putting it in a graph, we may want to begin with the prototype: $\csc x$.

First, setting the frequency to 2, we get: $\csc 2x$.

We know that the period is: the basic period (interval) over the frequency, and that the basic interval (period) for the cosecant is 2π. So we can see that the period is: $2\pi/2 = \pi$.

And thus, we can put the curve with $\csc 2x$ the way below:

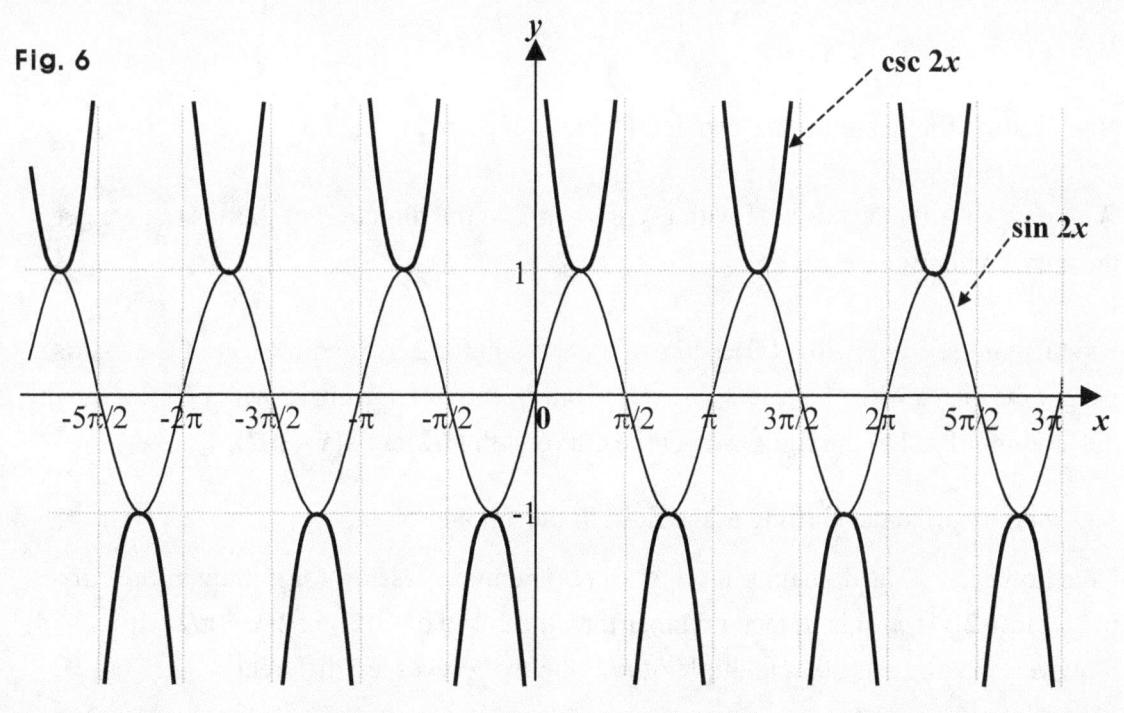

Fig. 6

Next, setting the constant A to 1/2, we get: **(1/2)csc 2x**.

And assuming $m(x) = $ **csc 2x** and $n(x) = $ **(1/2)csc 2x**, we get: $n(x) = $ **(1/2)m(x)** for every x-value. That is, for each and every x-value, the y-value in n is half the y-value in m.

And thus, we can put in a graph the curve with **(1/2)csc 2x** the way below:

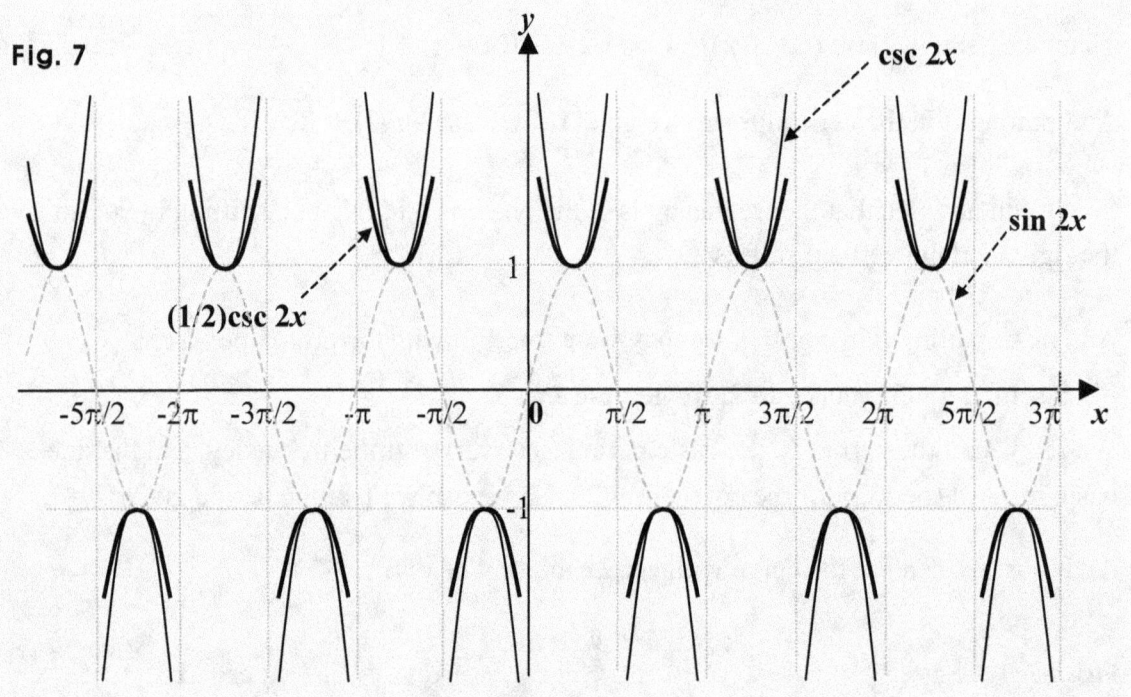

Fig. 7

Next, setting the phase to -π/2, we get: **(1/2)csc 2(x − π/2)**.

We know shifting the curve of with **csc x** by -π/2 in the direction of the x-axis, we get the curve with **csc (x + π/2)**.

So shifting the curve with **(1/2)csc 2x** by not -π/2 but π/2 in the direction of the x-axis, we get the curve with **(1/2)csc 2(x − π/2)**. That is, moving the curve with **(1/2)csc 2x** in the amount of π/2 to the right, we get the curve with **(1/2)csc 2(x − π/2)**.

So we do a horizontal shifting to the right in the amount of π/2.

Note however, if the domain is a set of all real numbers less $n\pi$ for n integer, the curve of **(1/2)csc 2(x − π/2)** is in fact, no other than the curve of **(1/2)csc 2(x + π/2)**. If however, the phase is not a multiple of **π/2**, the two curves are different.

And thus, we can put in a graph, the curve of **(1/2)csc 2(x − π/2)** the way below:

Fig. 8

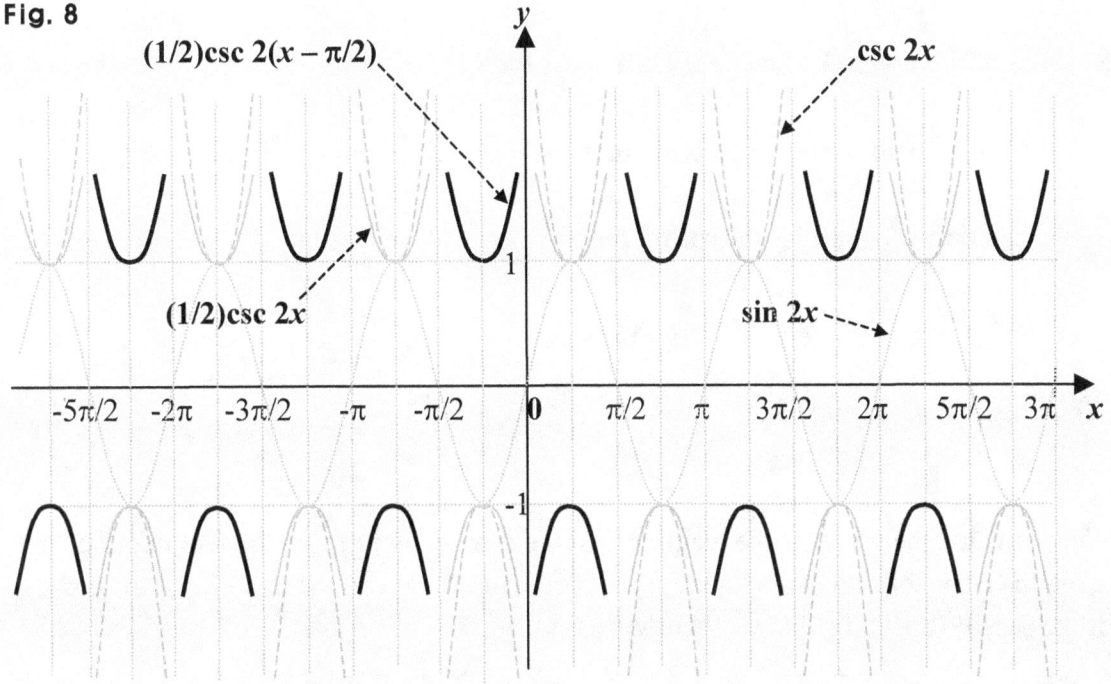

• And taking next, the <u>reciprocal</u> of the <u>cosine</u>, we get another trig-ratio called the <u>secant,</u> denoted by **sec**. And thus, taking the reciprocal of the cosine, that is, **cos x**, we get: **sec x**.

And we can take **cos x** as a function, set: $y = f(x) = \cos x$, and call it a cosine function. What then, about a secant function?

We can make another trig-function called a secant function using **sec x**, of course.
So for instance, we can make secant functions as $y = g(x) = \sec x$ or $y = q(x) = \sec 3x^2$.

As in the case of a cosecant function as $y = h(x) = \csc x$ however, the secant function above cannot be defined for all angles.   How come?

We know that <u>the cosine is: the adjacent over the hypotenuse.</u>
So <u>the secant is: the hypotenuse over the adjacent,</u> since it is the reciprocal of the cosine.
And we know that the adjacent can be 0 for some angles as -π/2 and π/2.

So for instance, the secant **sec** *x* cannot be defined for *x* = π/2. And in general, it cannot be defined for *x* = **(2n + 1)π/2** for *n* integer, that is, *x* = **nπ/2** where *n* is an odd integer.

So just setting: *y* = *g(x)* = **sec** *x*, we mean: *x* ≠ **nπ/2** for *n* odd.

That is, the domain is a set of all real numbers less **nπ/2** where *n* is an odd integer.

So we can define or specify the secant function the way below, too:

y = *g(x)* = **sec** *x* for *x* ≠ **nπ/2** for *n* odd.

What then, about its curve?

We know that the secant is the reciprocal of the cosine. So keeping the fact in mind, and using the curve of the cosine function *y* = *f(x)* = **cos** *x*, we can put in a graph, the curve of the secant function *g* the way below:

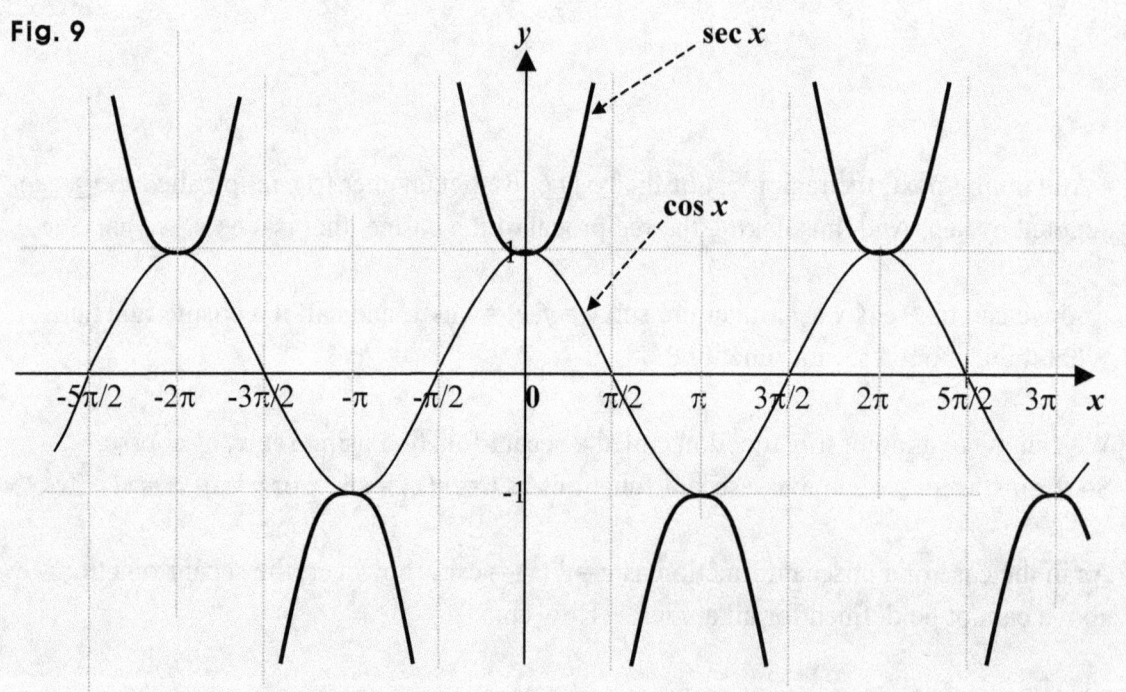

Fig. 9

So we can now see that the *y*-values, that is, the outputs are grater than or equal to 1 or less than or equal to –1. And thus, the range is: *y* ≤ **-1** or *y* ≥ **1**, that is, |*y*| ≥ **1**.

And of course, if the domain is different, the range can be other than the one above. Assuming for instance, $y = h(x) = \sec x$ for $0 < x < \pi/2$, we get: $y > 1$, which is the range.

And the secant function $y = g(x) = \sec x$ can be called the prototype, and is thus, in the most basic form. Assuming G is a secant function, too, and the domain is a set of all real numbers less $n\pi/2$ for n odd, and using a general form, we can put it the way as follows:

$y = G(x) = A \cdot \sec \{w(x + a)\} + b$ where $w(x + a) \neq n\pi/2$ for n odd, and A, w, a, and b are constant.

And we can just put it this way, too: $y = G(x) = A \cdot \sec w(x + a) + b$.

Then, $|w|$ is called the frequency, $\frac{2\pi}{|w|}$ is the period, and a is called the phase.

So in the function $y = g(x) = \sec x$, the phase is 0, and the frequency is 1.

And looking at the curve above, we can see that a part of the cuve repleats itself. And the part is in fact, the smallest part, in two pieces though, repeating itself.
What then, is the part?

It is the part from 0 to 2π. So the part repeats itself every 2π interval.

So the interval 2π is the period in the secant function $y = g(x) = \sec x$.

• And thus, we call a secant function a *periodic* function, too.

And the secant function g above is the prototype, and thus, is in the most basic form.

   • So the period 2π can be called the *basic period* in *secant functions*. In other words:
   • The interval 2π can be called the *basic interval* in *secant functions*.

And for instance, in the general form, setting w to 2, a and b to 0 each, and A to 1, we get a new function where: $y = k(x) = \sec 2x$. What then, is the domain?

We know that **sec θ** is not defined for $\theta = n\pi/2$ where *n* is an odd integer.

So in **sec 2x**, we don't want **2x** to be *nπ/2* for *n* odd.

That is, we want: **$2x \neq n\pi/2 \Rightarrow x \neq n\pi/4$** for *n* odd.

So the domain in the function **sec 2x** can be a set of all real numbers less *nπ/4* for *n* odd. And of course, if *E* is the domain just described, *E* is the biggest domain that the secant function **sec 2x** can have.

Then, putting in the *x-y* plane, the curve of the secant function *k(x)* = **sec 2x**, we get:

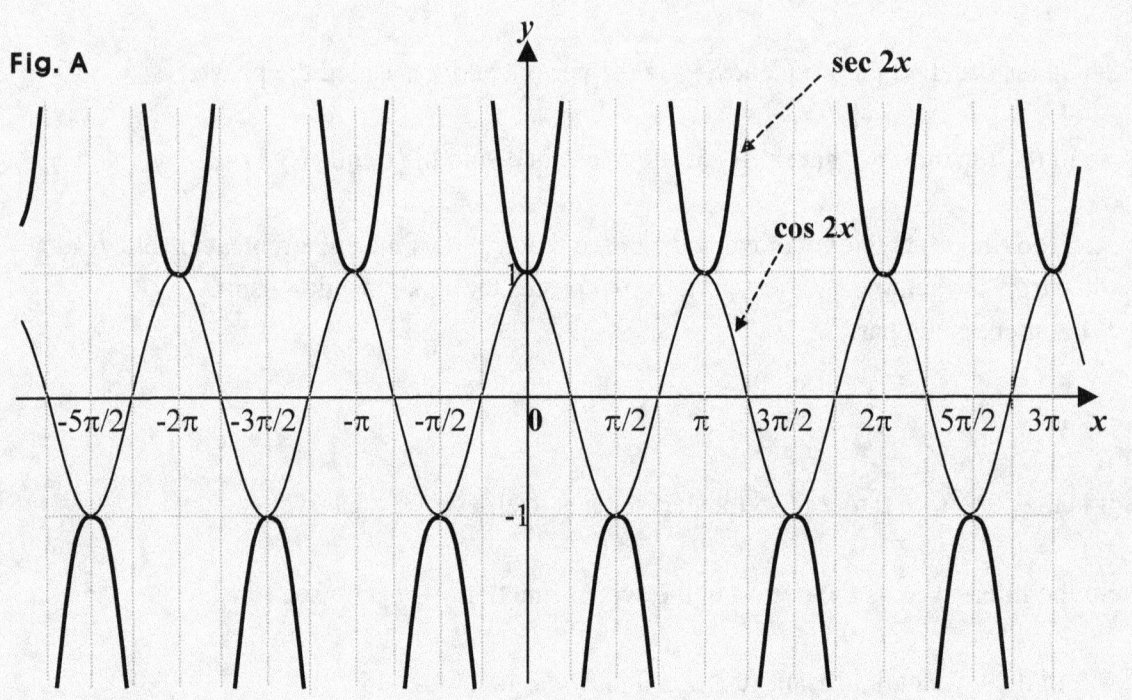

Fig. A

Then again, we can see that a part of the cuve repleats itself, and is in two pieces.

And the part is of course, the smallest part repeating itself.   What then, is the part?

It is the part from -π/2 to 3π/4. So the part repeats itself every π interval. And thus, the interval π is the period in the function *k*. And the smallest part stated above appears twice in the basic interval 2π, which is the basic period.

Then, we say the frequency is 2 in the function **k**.

• So in case of secant functions, the frequency is the number of times the samllest part repeating itself appears in the basic period or the basic interval 2π.

And thus, in the general form $A\cdot\sec w(x + a) + b$, $|w|$ is the frequency.

And we know that the interval that fits the smallest part is the period. So using the frequency $|w|$ and the basic interval 2π, we can put the period this way: $\frac{2\pi}{|w|}$.

And we can have a secant function where $y = h(x) = \sec(-x)$. Then, the frequency in the secant function **h** is $|-1|$, simply because a frequency is positive.

And we can put **h** this way, too: $y = h(x) = \sec x$, because $\sec(-x) = \sec x$. Why though?

We know the secant is: the hypotenuse over the adjacent.
So assuming the ray, that is, the hypotenuse is of length 1, we get:

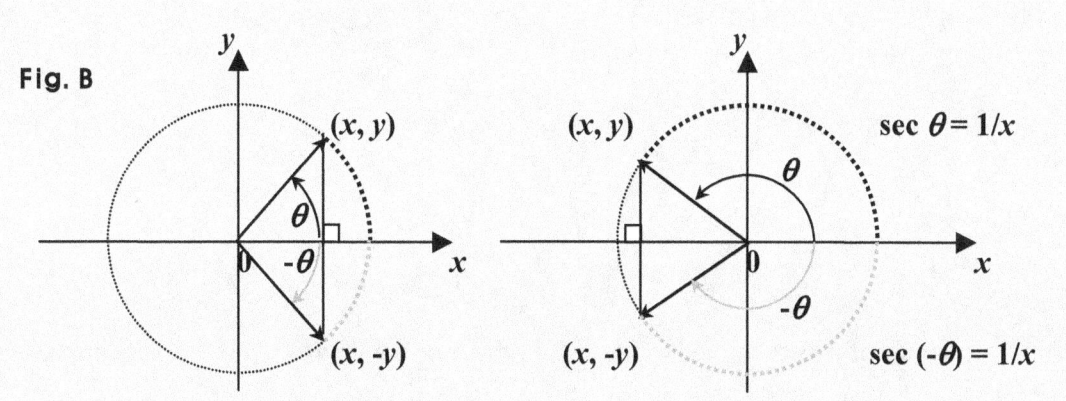

Fig. B

So we get: $\sec(-\theta) = \sec\theta$. And thus, putting in a graph, the curve of $y = h(x) = \sec(-x)$, we get the same curve as that of $y = g(x) = \sec x$, and the period of **h** is 2π, too.

Next, shifting by $-a$ in the direction of the **x**-axis, the curve of $y = Q(x) = A\cdot\sec wx + b$ where $wx \neq n\pi/2$ for **n** odd, we get the curve of a function below:

$y = G(x) = A\cdot\sec w(x + a) + b$ where $w(x + a) \neq n\pi/2$ for **n** odd.

So the two curves themselves of *G* and *Q* are the same, and moving the curve of *Q* in the amount of *–a* along the *x*-axis, we get the curve of *G*.

And we call *a* the phase, and if *a* is positive, the curve gets shifted to the left, and if negative, it moves to the right.

Assuming for instance, shifting the curve of $p(x) = \sec x$ for **-3π/2 < x < 5π/2** by –π/2 in the direction of the *x*-axis, that is, to the left, we get the curve of a function below:

$q(x) = \sec(x + \pi/2)$ for **-2π < x < 2π**.   So putting the two curves in one graph, we get:

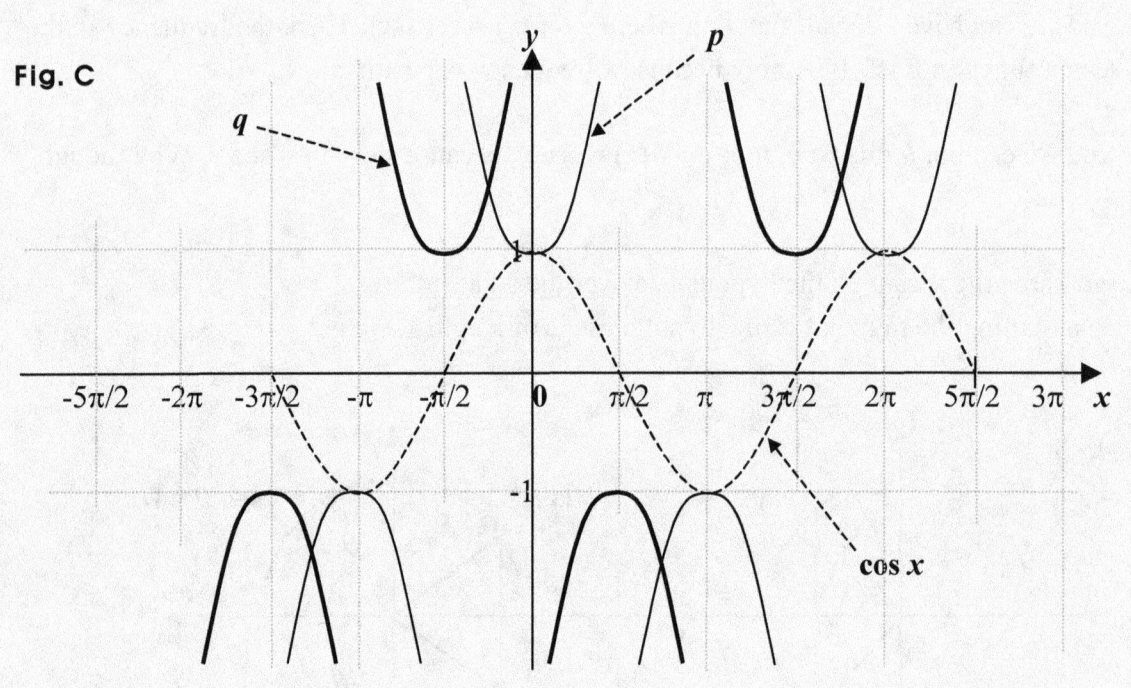

Fig. C

And thus, the curve of *p* gets moved to the left in the amount of π/2. And the new curve is the curve of *q*. So we can call the phase a horizontal shift, too.

What then, about the constant *b* in $y = Q(x) = A \cdot \sec wx + b$?

It is called a vertical shift.

So shifting the curve of $y = V(x) = A \cdot \sec w(x + a)$ where $w(x + a) \neq n\pi/2$ for *n* odd, by *b* in the direction of the *y*-axis, we get the curve of the function *G* as follows:

$y = G(x) = A \cdot \sec w(x + a) + b$ where $w(x + a) \neq n\pi/2$ for *n* odd.

So the two curves themselves of *G* and *V* are the same, and moving the curve of *V* in the amount of *b* along the *y*-axis, we get the curve of *G*.

And if *b* is positive, the curve is shifted upward, and if negative, it moves downward.

Assuming for instance, shifting by 1/2 in the direction of the *y*-axis, that is, upward, the curve of *u(x)* = sec *x*, we get the curve of a function as follows: *v(x)* = sec *x* + 1/2.

Putting therefore, the two curves in a graph, we get:

Fig. D

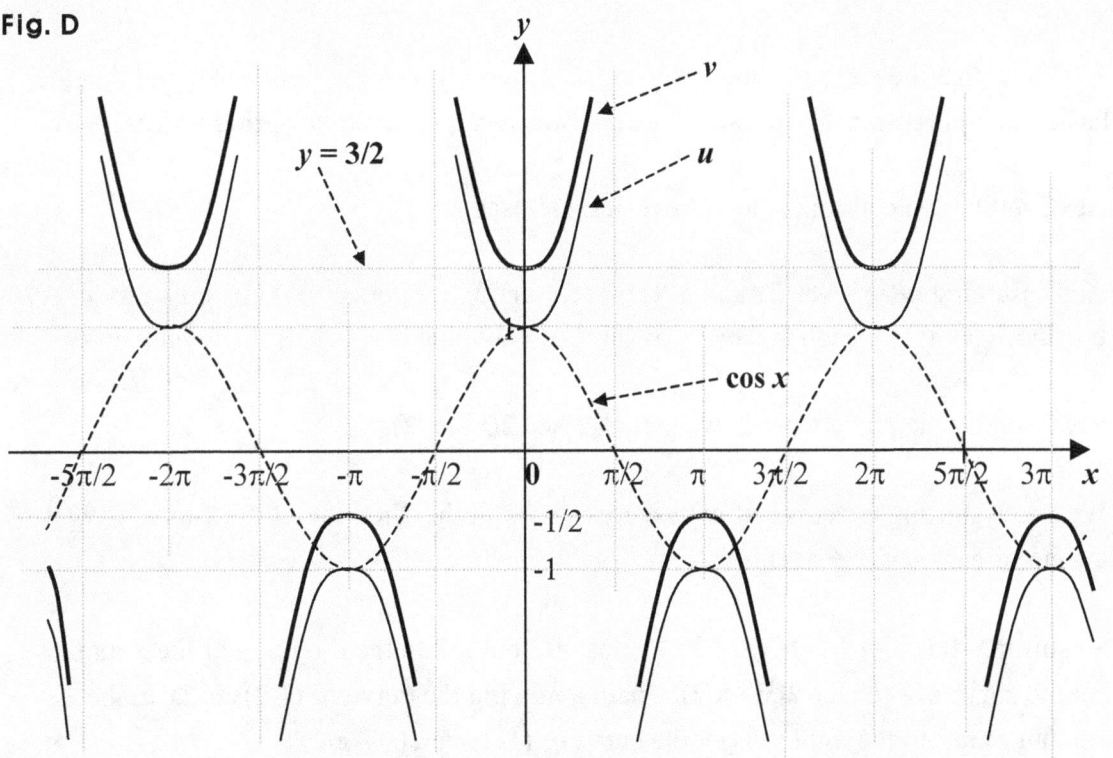

So the curve of *u* gets moved upward in the amount of 1/2. And the new curve is the curve of *v*. So we can call the *b* a vertical shift.

Suppose now, we put in a graph, for instance, the curve of a secant function below:

y = *S(x)* = (-1/2)sec (-2*x* + π) + 1, where -2*x* + π ≠ *n*π/2 for *n* odd.

Then, to begin with, we have: **sec (-*θ*) = sec *θ*.**
So we can get: **sec (-2*x* + π) = sec {-(2*x* − π)} = sec (2*x* − π).**

Thus, we get: **(-1/2)sec (-2*x* + π) + 1 = (-1/2)sec (2*x* – π) + 1.**

And putting it in the general form, we get: **(-1/2)sec 2(*x* – π/2) + 1.**

So we can now see that the frequency is 2, the phase is –π/2, that is, shifting by π/2 to the right, and the vertical shift is 1.

And next, putting it in a graph, we may want to begin with the prototype: **sec *x*.**

First, setting the frequency to 2, we get: **sec 2*x*.**

We know that the period is: the basic period (interval) over the frequency, and that the basic interval (period) for the secant is 2π. So we can see that the period is: 2π/2 = π.

Next, setting the constant *A* to 1/2, we get: **(1/2)sec 2*x*.**

And assuming *m*(*x*) = sec 2*x* and *n*(*x*) = (1/2)sec 2*x*, we get: *n*(*x*) = (1/2)*m*(*x*) for every *x*-value. That is, for each and every *x*-value, the *y*-value in *n* is half the *y*-value in *m*.

Next, setting the phase to -π/2, we get: **(1/2)sec 2(*x* – π/2).**

We know shifting the curve of with **sec *x*** by -π/2 in the direction of the *x*-axis, we get the curve with **sec (*x* + π/2).**

So shifting the curve of **(1/2)sec 2*x*** by not -π/2 but π/2 in the direction of the *x*-axis, we get the curve of **(1/2)sec 2(*x* – π/2).** That is, moving the curve of **(1/2)sec 2*x*** in the amount of π/2 to the right, we get the curve of **(1/2)sec 2(*x* – π/2).**

So we do a horizontal shifting to the right in the amount of π/2.

Note however, if the domain is a set of all real numbers less *n*π/2 for *n* odd, the curve of **(1/2)sec 2(*x* – π/2)** is in fact, no other than the curve of **(1/2)sec 2(*x* + π/2).** If however, the phase is not a multiple of **π/2**, the two curves are different.

- And taking next, the reciprocal of the tangent, what do we get?

We get another trig-ratio called the cotangent, denoted by **cot**.
So taking the reciprocal of the tangent, that is, **tan** x, we get: **cot** x.

And taking **tan** x as a function, we can set: $y = f(x) = $ **tan** x, and call it a tangent function.

So we can make another trig-function called a cotangent function using **cot** x, of course.
For instance, we can make a cotangent function as $y = g(x) = $ **cot** x or $y = h(x) = $ **2cot** x^3.

However, the cotangent function **cot** x cannot be defined for all angles. How come?

We know that <u>the tangent is: the opposite over the adjacent</u>.
So <u>the cotangent is: the adjacent over the opposite</u> since it is the reciprocal of the tangent.

And we know that the opposite can be 0 for some angles as 0, -π, and π. So for instance,
the cotangent **cot** x cannot be defined for $x = π$. And in general, it cannot be defined for
$x = 2n\pi$ where n is an integer. That's not all though. What else then?

Unlike the tangent, taking the cotangent, **cot** x, we take not just the ratio of the adjacent
to the opposite but the reciprocal of the tangent, **tan** x, too.

So taking **cot** x, we want to make sure that the reciprocal of **tan** x is defined.

And in fact, the same is true, too, for **csc** x and **sec** x, the reciprocals of **sin** x and **cos** x.
Why though?

To begin with, we know **sin** $x = 0$ for $x = n\pi$ where n is an integer.
So for $x = n\pi$ where n is an integer, we get: **csc** $x = $ 1/**sin** $x = $ 1/0, which is not allowed.

And thus, **csc** x cannot be defined for $x = n\pi$ where n is an integer.
In fact, for $x = n\pi$ where n is an integer, the opposite is 0, so **csc** x cannot be defined.

So in sum, **csc** x cannot be defined for $x = n\pi$ where n is an integer.

Next, we know **cos** $x = 0$ for $x = n\pi/2$ where n is an odd integer.

So for $x = n\pi/2$ for n odd, we get: **sec** $x = 1/\cos x = 1/0$, which is not allowed. And thus, **sec** x cannot be defined for $x = n\pi/2$ where n is an integer odd.

In fact, for $x = n\pi/2$ where n is an odd integer, the adjacent is 0, so **sec** x cannot be defined. So in sum, **sec** x cannot be defined for $x = n\pi/2$ for n odd.

What then, about **cot** x?

We know: **tan** $x = 0$ for $x = n\pi/2$ where n is an odd integer.

So for $x = n\pi/2$ for n odd, we get: **cot** $x = 1/\tan x = 1/0$, which is not allowed. And thus, **cot** x cannot be defined for $x = n\pi/2$ where n is an integer odd.

And also, we know that **cot** x cannot be defined either for $x = 2n\pi$ for n integer.

So in sum, the cotangent **cot** x cannot be defined for $x = 2n\pi$ or $(2n + 1)\pi/2$ for n integer.

Thus, just setting: $y = g(x) = \cot x$, we mean that $x \neq 2n\pi$ or $(2n + 1)\pi/2$ for n integer.

That is, the domain is a set of all real numbers less $2n\pi$ and $(2n + 1)\pi/2$ for n integer.

So we can define or specify the cotangent function the way below, too:

$y = g(x) = \cot x$ for $x \neq 2n\pi$ or $(2n + 1)\pi/2$ for n integer.

What then, about its curve?

We know that the cotangent is the reciprocal of the tangent. So keeping the fact in mind, and using the curve of the tangent function $y = f(x) = \tan x$, we can put in a graph, the curve of the cotangent function $y = g(x) = \cot x$ the way below:

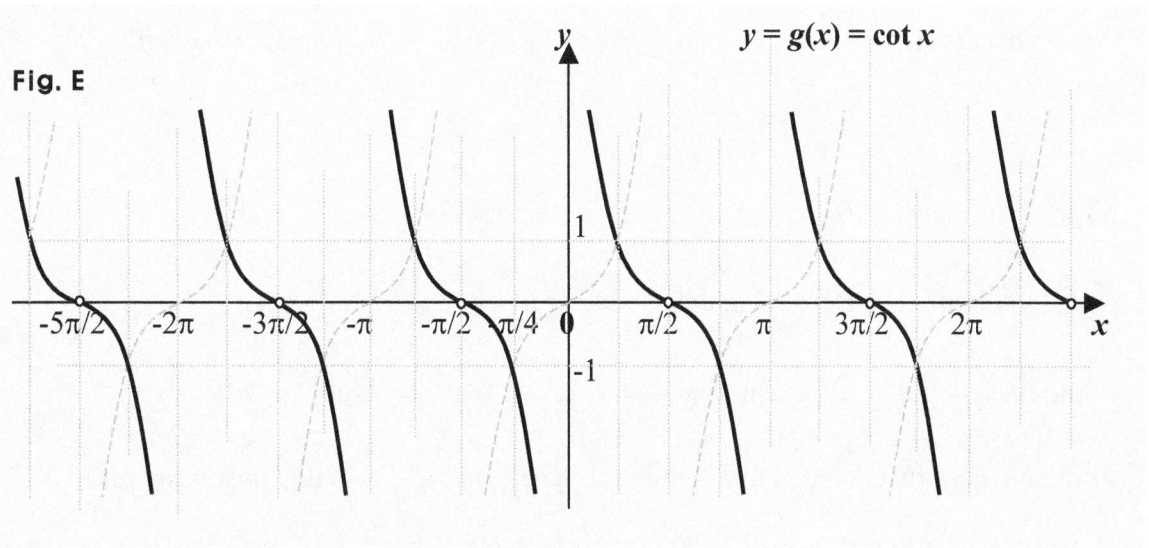

Note that there is a point missing at every $n\pi/2$ for n odd.

It's because the cotangent function g is *not* defined not only for $x = 2n\pi$ for n integer but for $x = n\pi/2$ for n odd, too.

So we can now see that the y-values, that is, the outputs are all real numbers other than 0. And thus, the range is: $y \neq 0$.

And of course, if the domain is different, the range can be other than the one above. Assuming for instance, $\underline{y = h(x) = \cot x \text{ for } 0 < x < \pi/2}$, we get: $y > 0$, which is the range.

And the cotangent function $y = g(x) = \cot x$ can be called the prototype.
Assuming G is a cotangent function, too, and the domain is a set of all real numbers less $2n\pi$ and $(2n + 1)\pi/2$ for n integer, and using a general form, we can put it this way:

$y = G(x) = A \cdot \cot \{w(x + a)\} + b$ where $w(x + a) \neq 2n\pi$ or $(2n + 1)\pi/2$ for n integer, and A, w, a, and b are constant.

And just simply, we can put it this way, too: $y = G(x) = A \cdot \cot w(x + a) + b$.

Then, $|w|$ is called the frequency, $\frac{\pi}{|w|}$ is the period, and a is called the phase.

So in the function $y = g(x) = \cot x$, the phase is 0, and the frequency is 1. And looking at the curve above, we can see that a part of the cuve repleats itself. And the part is in fact, the smallest part, in two pieces though, repeating itself.   What then, is the part?

It is the part between 0 and π. So the part repeats itself every π interval.

So the interval π is the period in the cotangent function $y = g(x) = \cot x$.

• And thus, we call a cotangent function a *periodic* function, too.

And the secant function g above is the prototype, and thus, is in the most basic form.

• So the period π can be called the *basic period* in *cotangent functions*. In other words:

• The interval π can be called the *basic interval* in *cotangent functions*.

And for instance, in the general form, setting w to 2, a and b to 0 each, and A to 1, we get a new function where: $y = k(x) = \cot 2x$.   What is the domain though?

We know that $\cot \theta$ is not defined for $\theta = 2n\pi$ and $(2n + 1)\pi/2$ where n is an integer.

So in $\cot 2x$, we don't want $2x$ to be $2n\pi$ or $(2n + 1)\pi/2$ for n integer.

That is, we want: $2x \neq 2n\pi$ or $(2n + 1)\pi/2 \Rightarrow x \neq n\pi$ or $(2n + 1)\pi/4$ for n integer.

And thus, the domain in the function $k(x) = \cot 2x$ can be a set of all real numbers less $n\pi$ and $(2n + 1)\pi/4$ for n integer.

And of course, if E is the domain just described, E is the biggest domain that the cotangent function $\cot 2x$ can have.

Then, putting in the *x-y* plane, the curve of the cotangent function $k(x) = \cot 2x$, we get:

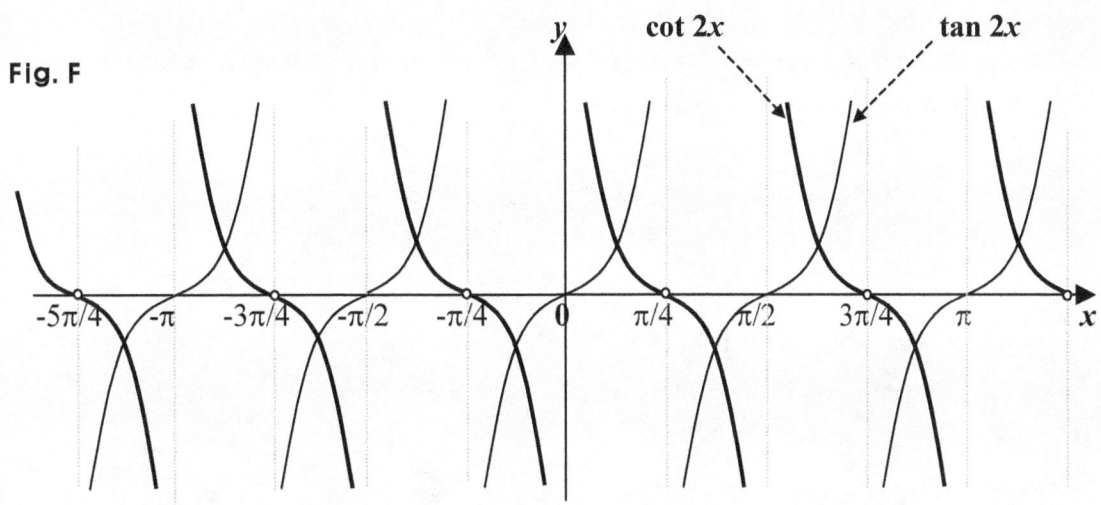

Fig. F

Then again, we can see that a part of the cuve repleats itself, and is in two pieces.

And the part is of course, the smallest part repeating itself.   What then, is the part?

It is the part from 0 to $\pi/2$. So the part repeats itself every $\pi/2$ interval. And thus, the interval $\pi/2$ is the period in the function $k(x) = \cot 2x$. And the smallest part stated above appears twice in the basic interval π, which is the basic period.

So we say the frequency is 2 in the function *k*.

• So in case of cotangent functions, the frequency is the number of times the samllest part repeating itself appears in the basic period or the basic interval π.

And thus, in the general form $A \cdot \cot w(x + a) + b$, $|w|$ is the frequency.

And we know that the interval that fits the smallest part is the period. So using the frequency $|w|$ and the basic interval π, we can put the period this way: $\frac{\pi}{|w|}$.

And thus, the period of the function $y = k(x) = \cot 2x$ is: $\pi/2$.

And we can have a cotangent function $y = h(x) = \cot(-x)$ for $x \neq n\pi$ or $(2n + 1)\pi/4$ for n integer.

Then, the frequency in the cotangent function h is $|-1|$, simply because a frequency is positive.

And we can put h this way, too: $y = h(x) = -\cot x$ for $x \neq n\pi$ or $(2n + 1)\pi/4$ for n integer. It's because $\cot(-x) = -\cot x$.   How come?

We know <u>the cotangent is: the adjacent over the opposite.</u>
So assuming the ray, that is, the hypotenuse is of length 1, we get:

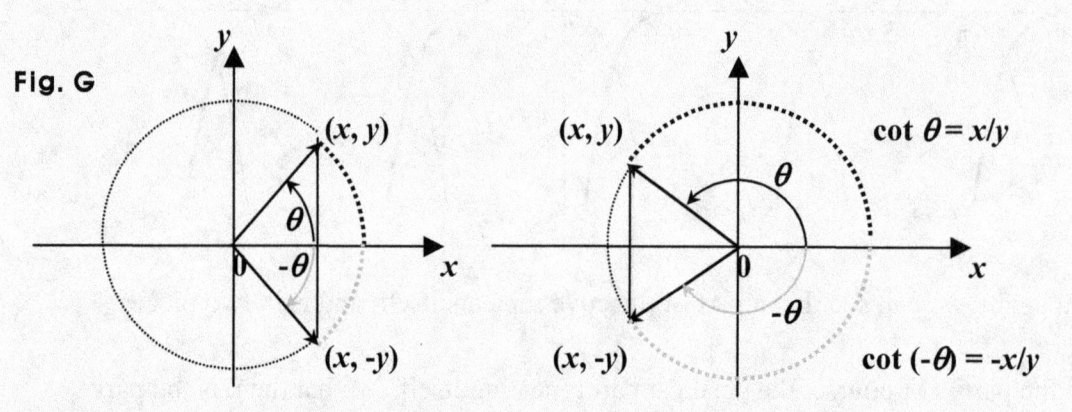

Fig. G

So we get: $\cot(-\theta) = -\cot \theta$.
And thus, putting in a graph, the curve of $y = h(x) = \cot(-x)$, we get the curve as below:

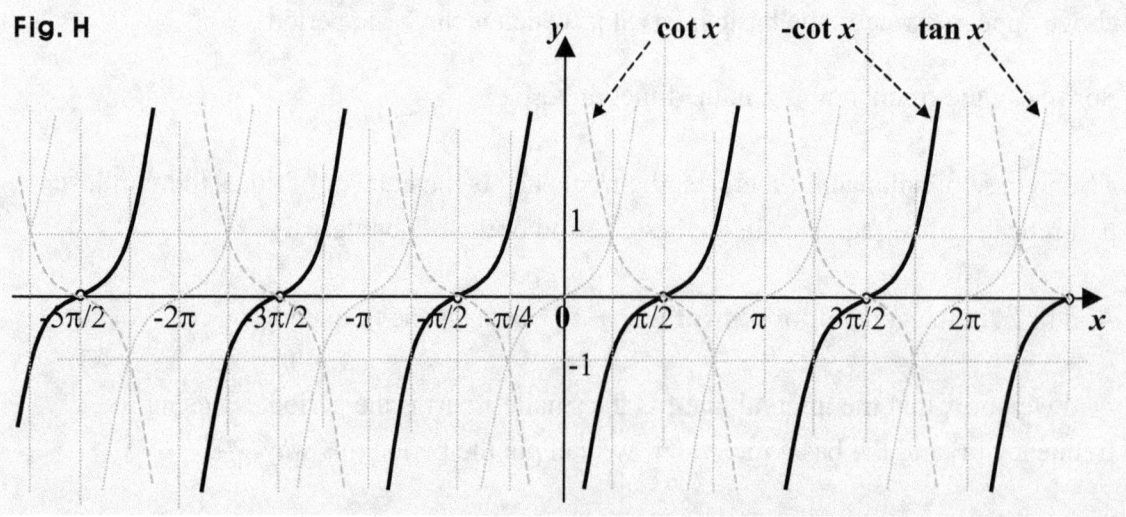

Fig. H

And we can see that the period of $h(x) = \cot(-x)$ is 2π, too.

Next, shifting by −*a* in the direction of the *x*-axis, the curve of *y* = *Q*(*x*) = *A*·cot *wx* + *b*, we get the curve of *y* = *G*(*x*) = *A*·cot *w*(*x* + *a*) + *b*.

So the two curves themselves of **G** and **Q** are the same, and moving the curve of **Q** in the amount of −*a* along the *x*-axis, we get the curve of **G**.

And we call *a* the phase, and if *a* is positive, the curve gets shifted to the left, and if negative, it moves to the right.

Assuming for instance, shifting the curve of *p*(*x*) = cot *x* for -π < *x* < 2π by −π/2 in the direction of the *x*-axis, that is, to the left, we get the curve of a function below:

q(*x*) = cot (*x* + π/2) for -2π < *x* < 2π.   So putting the two curves in a graph, we get:

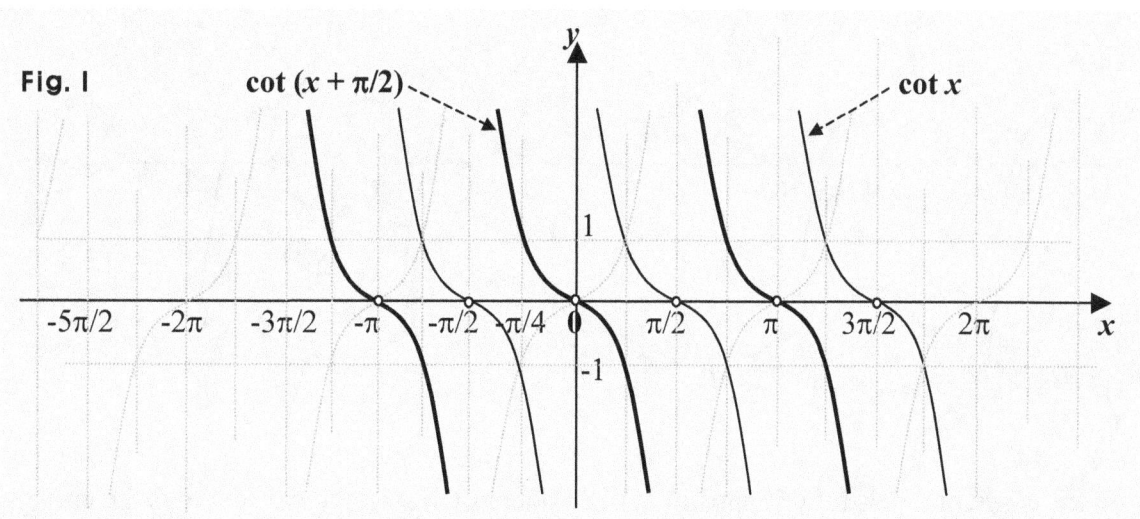

And thus, the curve of *p* gets moved to the left in the amount of π/2. And the new curve is the curve of *q*. So we can call the phase a horizontal shifting, too.

  • What then, about the constant *b* in *y* = *Q*(*x*) = *A*·cot *wx* + *b*?

It can be called a vertical shift.

So shifting the curve of *y* = *V*(*x*) = *A*·cot *w*(*x* + *a*) by *b* in the direction of the *y*-axis, we get the curve of *y* = *G*(*x*) = *A*·cot *w*(*x* + *a*) + *b*.

So the two curves themselves of **G** and **V** are the same, and moving the curve of **V** in the amount of **b** along the **y**-axis, we get the curve of **G**.

And if **b** is positive, the curve is shifted upward, and if negative, it moves downward.

Assuming for instance, shifting by 1 in the direction of the **y**-axis, that is, upward, the curve of **u(x)** = **cot x**, we get the curve of a function as follows: **v(x)** = **cot x + 1**.

Putting therefore, the two curves in a graph, we get:

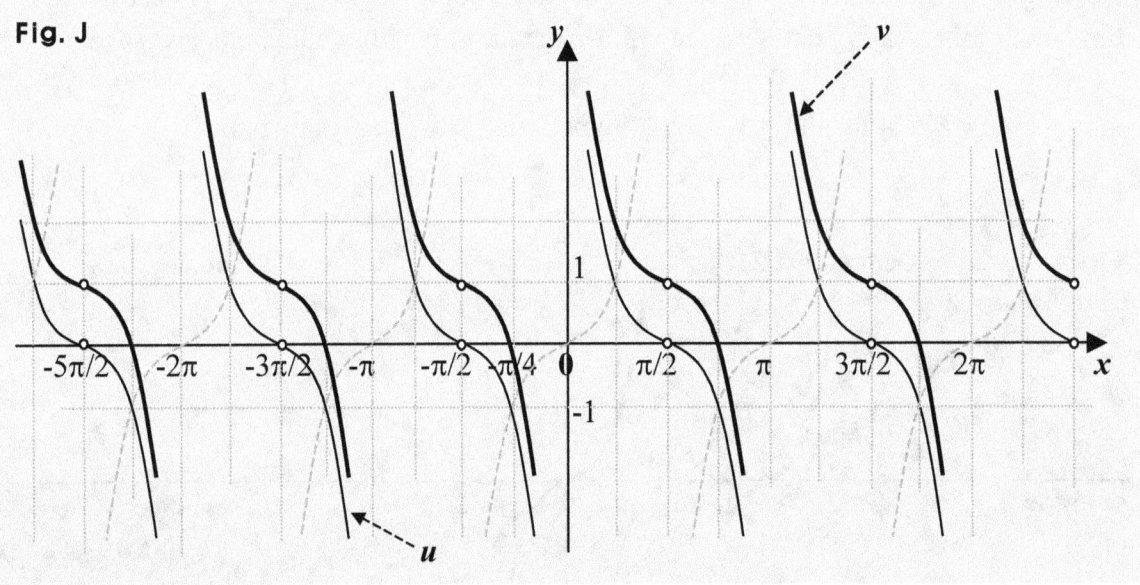

Fig. J

So the curve of **u** gets moved upward in the amount of 1. And the new curve is the curve of **v**. So we can call the **b** a vertical shift.

And let's now put in a graph, for instance, the curve of a cotangent function below:

y = **S(x)** = **-2cot (-2x + π) + 1**, where **-2x + π ≠ nπ** or **(2n + 1)π/4** for **n** integer.

Then, to begin with, we have: **cot (-θ) = -cot θ**.

So we can get: **cot (-2x + π) = cot {-(2x − π)} = -cot (2x − π)**.

Thus, we get: **-2cot (-2x + π) + 1 = 2cot (2x − π) + 1**.

And putting it in the general form, we get: **2cot 2(x – π/2) + 1**. How come?

Note that putting a cotangent function in general, we can put it the way below:

y = *G*(*x*) = *A*·cot *w*(*x* + *a*) + *b*.

Then, |*w*| is called the frequency, $\frac{\pi}{|w|}$ is the period, and *a* is called the phase.

So we can now see that the frequency is 2, the phase is –π/2, that is, shifting by π/2 to the right, and the vertical shift is 1.

And next, putting it in a graph, we may want to begin with the prototype: **cot x**.

First, setting the frequency to 2, we get: **cot 2x**.

We know that the period is: the basic period (interval) over the frequency, and that the basic interval (period) for the cotangent is π. So we can see that the period is: π/2.

Next, setting the constant *A* to 2, we get: **2cot 2x**.

And assuming *m*(*x*) = **cot 2x** and *n*(*x*) = **2cot 2x**, we get: *n*(*x*) = 2*m*(*x*) for every *x*-value.

That is, for each and every *x*-value, the *y*-value in *n* is twice the *y*-value in *m*.

Next, setting the phase to -π/2, we get: **2cot 2(x – π/2)**.

We know shifting the curve of with **cot x** by -π/2 in the direction of the *x*-axis, we get the curve with **cot (x + π/2)**.

So shifting the curve with **2cot 2x** by not -π/2 but π/2 in the direction of the *x*-axis, we get the curve with **2cot 2(x – π/2)**. That is, moving the curve with **2cot 2x** in the amount of π/2 to the right, we get the curve with **2cot 2(x – π/2)**.

So we do a horizontal shifting to the right in the amount of $\pi/2$.

Note however, if the domain is a set of all real numbers less **$n\pi$** or **$(2n + 1)\pi/4$** for **n** integer, the curve of **$2\cot 2(x - \pi/2)$** is in fact, no other than the curve of **$2\cot 2(x + \pi/2)$**. If however, the phase is not a multiple of **$\pi/2$**, the two curves are different.

E. The Inverse Functions

The inverse of a function is called an inverse function, and is often just called the inverse.
What then, is an inverse function?

If a function gets an input, it produces an output, which is the output for the input.
For instance, putting an input 1 into the input variable x in a function $y = f(x) = 2x + 1$,
we get: $f(1) = 2 + 1 = 3$, which is called an output, which is the output for the input 1.

What then, about the inverse?

The inverse of a function is a function that produces an input for an output, which is the
output for the input.
So for instance, assuming g is the inverse of f, we get: $g(3) = 1$, because $f(1) = 3$.
Thus, we can say that an inverse function is a function that produces the set of inputs
using the set of inputs.

  • In short, the inverse produces the inputs using the outputs.

How then, can we get the inverse?

There are two conditions.

One is that the inverse produces the domain using the range.
In other words, the domain of the inverse is the range of the function we take the inverse
of, and the range of the inverse is the domain of the function we take the inverse of.
So the domain and the range are switched, which is however, a necessary condition, and
is not a sufficient condition.
It's because not every function where the domain and range are switched is the inverse.

What then, is the other condition?

The inverse is a function, too.
So the function we take the inverse of has to be a function one-to-one.
In short, the original function has to be one-to-one.   Why though?

First, there is no function that is one-to-many.
So next, if the original is not one-to-one, that is, many-to-one, the inverse would be one-to-many, and thus, is not a function. So no inverse function exists.

And the same is true for the inverse of a trig-function, too.

We know that the domain of a trig-function as **sin x** is a set of angles, and the range is a set of trig-ratios.
That is, each input of an inverse function is not an angle but a trig-ratio, and each output is not a trig-ratio but an angle.
So the domain of the inverse is the set of trig-ratios, and the range is the set of angles.
And of course, the trig-function we take the inverse of has to be one-to-one.
So what is the inverse of a trig-function?

Math begins with definitions. So we may want to begin with some definitions.
We can define the inverses of the three basic trig-functions the way as follows:

$$y = f(x) = \sin x \Leftrightarrow x = f^{-1}(y) = \sin^{-1} y,$$

$$y = g(x) = \cos x \Leftrightarrow x = g^{-1}(y) = \cos^{-1} y, \text{ and}$$

$$y = h(x) = \tan x \Leftrightarrow x = h^{-1}(y) = \tan^{-1} y, \text{ where } x \text{ is an angle, and } y \text{ is a trig-ratio.}$$

And specifying a particular inverse function, we usually use x as the input variable, and use y as the output variable. So for instance:

To begin with, a sine function where $y = f(x) = \sin x$ for $-\pi/2 \leq x \leq \pi/2$ is one-to-one.

We know the domain of f is: $-\pi/2 \leq x \leq \pi/2$, and the range is: $|y| \leq 1$.

So assuming g is the inverse of the sine function f, we get: $y = g(x) = \sin^{-1} x$ for $|x| \leq 1$, where x is a trig-ratio, and y is an angle. And the range of g is: $-\pi/2 \leq y \leq \pi/2$.

And we can call g an arc sine function, too, since $\sin^{-1} x$ is called the arc sine of x. What then, is the curve of the arc sine function g, which is the inverse of the sine function $y = f(x) = \sin x$ for $-\pi/2 \leq x \leq \pi/2$?

The curve of the inverse of a function is symmetric to the curve of the function about the line $y = x$. So the curve of g is symmetric to the curve of f about the line $y = x$, too. And thus, putting both the two functions f and g in a graph, we get:

Fig. 0

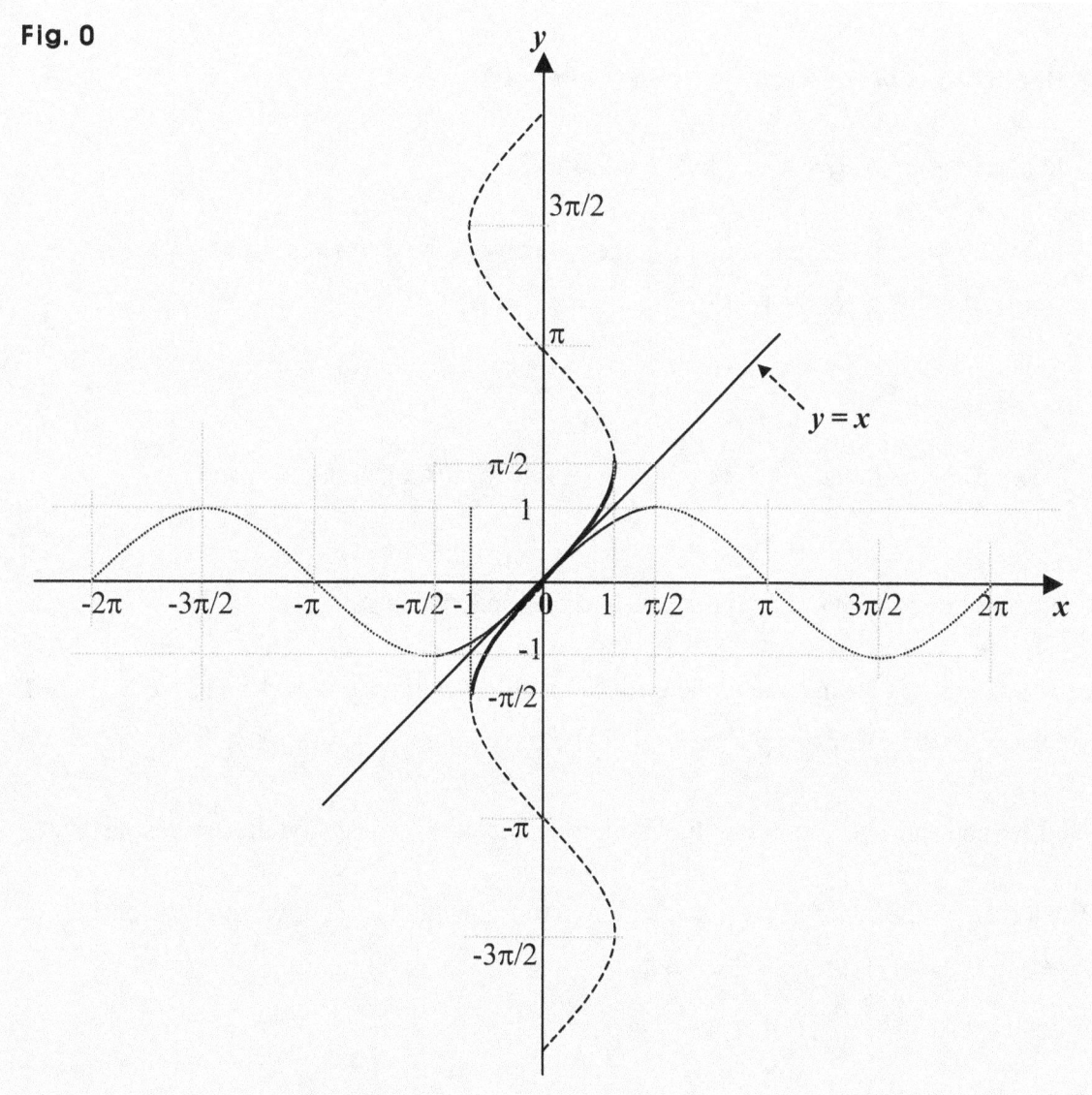

So for instance, $g(1) = \sin^{-1} 1 = \pi/2$.

And $g(-1) = \sin^{-1} (-1) = -\pi/2$.

What then, about the inverse of another sine function $y = p(x) = \sin x$ for $\pi/2 \le x \le 3\pi/2$?

We know $\sin^{-1}(-1) = -\pi/2$, and $\sin^{-1} 1 = \pi/2$.

So assuming q is the inverse, we can put q the way below:

$y = q(x) = -\sin^{-1} x + \pi$ for $|x| \le 1$.

So for instance, $p(1) = -\sin^{-1} 1 + \pi = -\pi/2 + \pi = \pi/2$.

And $p(-1) = -\sin^{-1}(-1) + \pi = -(-\pi/2) + \pi = 3\pi/2$.

And by the same token, assuming v is the inverse of $y = u(x) = \sin x$ for $-3\pi/2 \le x \le -\pi/2$, we can put v this way: $y = v(x) = -(\sin^{-1} x + \pi)$ for $|x| \le 1$.

Next, a cosine function where $y = f(x) = \cos x$ for $0 \le x \le \pi$ is one-to-one.

We know the domain in f is: $0 \le x \le \pi$, and the range is: $|y| \le 1$.

So assuming g is the inverse of the cosine function f, we get: $y = g(x) = \cos^{-1} x$ for $|x| \le 1$, where x is a trig-ratio, and y is an angle. And the range of g is: $0 \le y \le \pi$.

And we can call g an arc cosine function, too, since $\cos^{-1} x$ is called the arc cosine of x.

What then, is the curve of the arc cosine function g, which is the inverse of the cosine function $y = f(x) = \cos x$ for $0 \le x \le \pi$?

The curve of the inverse of a function is symmetric to the curve of the function about the line $y = x$.

So the curve of g is symmetric to the curve of f about the line $y = x$, too.

And thus, putting both the two functions f and g in a graph, we get:

Fig. 1

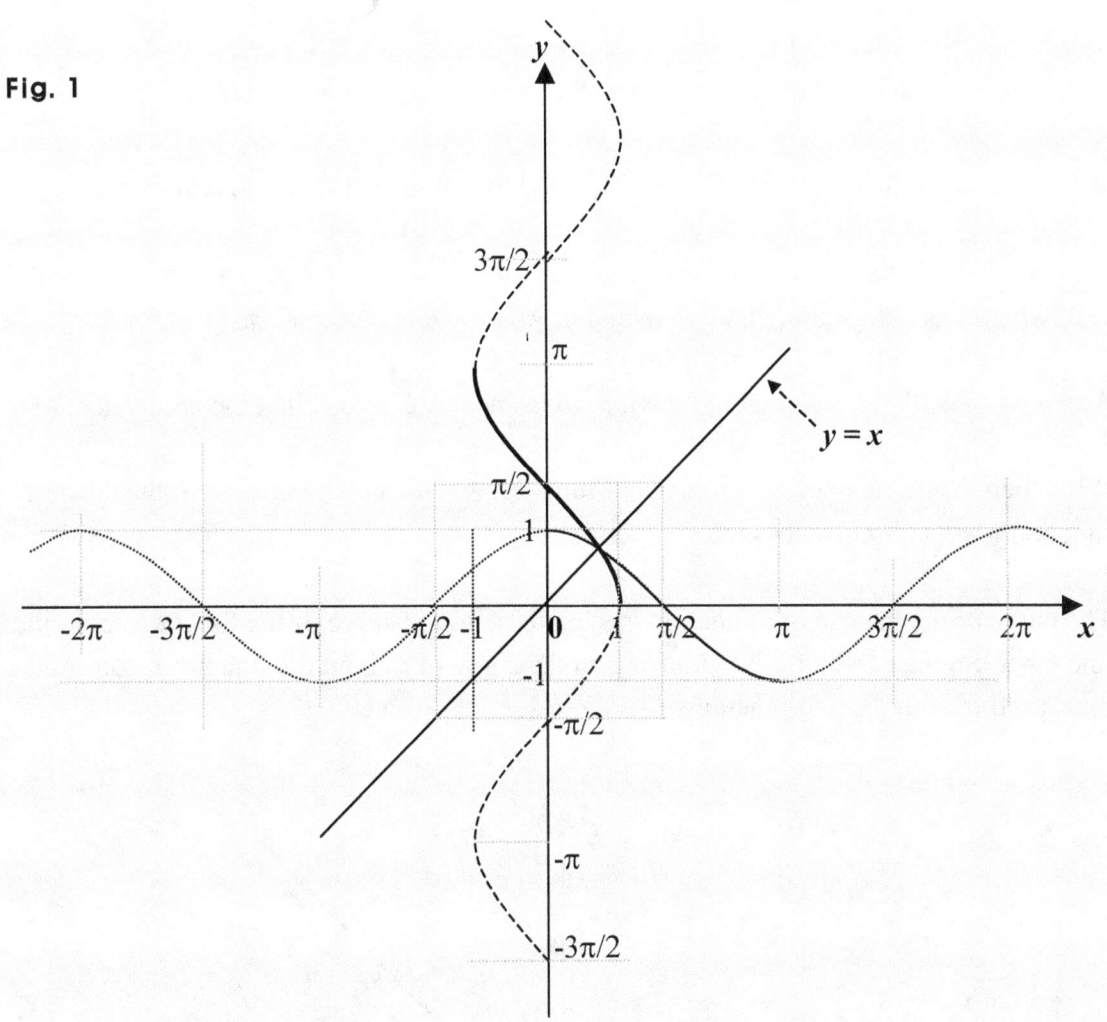

So for instance, $g(1) = \cos^{-1} 1 = 0$.   And $g(-1) = \cos^{-1}(-1) = \pi$.

What then, about the inverse of another cosine function $y = p(x) = \cos x$ for $\pi \le x \le 2\pi$?

We know $\cos^{-1}(-1) = \pi$, and $\cos^{-1} 1 = 0$.

So assuming q is the inverse, we can put q the way below:

$y = q(x) = \cos^{-1} x + \pi$ for $|x| \le 1$.

So for instance, $p(1) = \cos^{-1} 1 + \pi = 0 + \pi = \pi$. And $p(-1) = \cos^{-1}(-1) + \pi = \pi + \pi = 2\pi$.

And by the same token, assuming v is the inverse of $y = u(x) = \cos x$ for $-\pi \leq x \leq 0$, we can put v this way: $y = v(x) = -\cos^{-1} x$ for $|x| \leq 1$.

And next, a tangent function where $y = f(x) = \tan x$ for $-\pi/2 < x < \pi/2$ is one-to-one.

We know the domain in f is: $-\pi/2 < x < \pi/2$, and the range is a set of all real numbers.

So assuming g is the inverse of the tangent function f, we get: $y = g(x) = \tan^{-1} x$ for x real, where x is a trig-ratio, and y is an angle. And the range of g is: $-\pi/2 < y < \pi/2$.

And we can call g an arc tangent function, too, since $\tan^{-1} x$ is called the arc tangent of x.

What then, is the curve of the arc tangent function g, which is the inverse of the tangent function $y = f(x) = \tan x$ for $-\pi/2 < x < \pi/2$?

The curve of the inverse of a function is symmetric to the curve of the function about the line $y = x$. So the curve of g is symmetric to the curve of f about the line $y = x$, too. And thus, putting both the two functions f and g in a graph, we get:

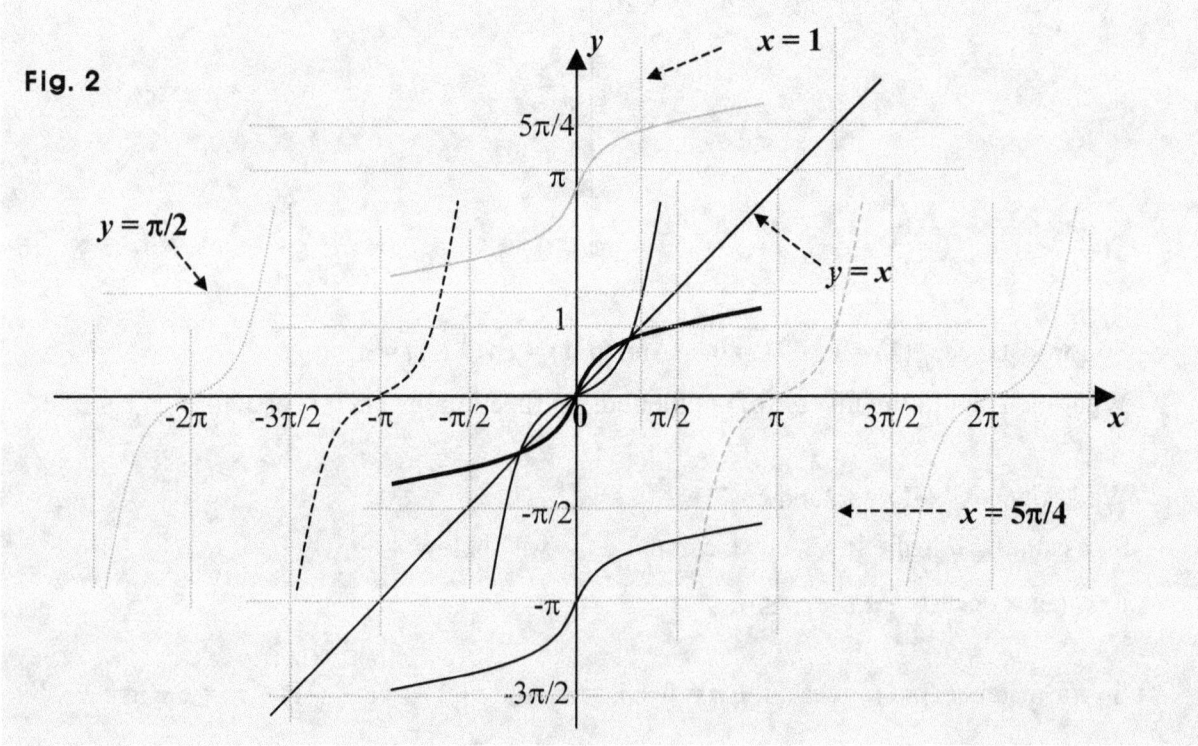

Fig. 2

So for instance, $g(1) = \tan^{-1} 1 = \pi/4$. And $g(-1) = \tan^{-1}(-1) = -\pi/4$.

What then, about the inverse of a tangent function $y = p(x) = \tan x$ for $\pi/2 < x < 3\pi/2$?

We know $\tan^{-1}(-1) = -\pi/4$, and $\tan^{-1} 1 = \pi/4$. So assuming q is the inverse, we can put q the way as follows: $y = q(x) = \tan^{-1} x + \pi$ for x real.

So for instance, $p(1) = \cos^{-1} 1 + \pi = \pi/4 + \pi = 5\pi/4$.

And $p(-1) = \tan^{-1}(-1) + \pi = -\pi/4 + \pi = 3\pi/4$.

And by the same token, assuming v is the inverse of $y = u(x) = \tan x$ for $-3\pi/2 \le x \le -\pi/2$, we can put v this way: $y = v(x) = \tan^{-1} x - \pi$ for x real.

F. Samples on Trig—Algebra

Note: $(\sin x)(\cos x) = \sin x \cos x$, and $(\sin x)^n = \sin^n x$, so for instance, $(\sin x)^2 = \sin^2 x$.

Solving problems, we get to do algebra.

So solving problems with trigonometry, we need to do algebra, too.
And doing algebra, we often get to work with expressions with trig-ratios as $\sin x$.
We can called such an expression a trig-expression. And doing algebra with such expressions, we can say we do trig-algebra.

Doing trig-algebra, we manipulate trig-expressions so that we can change them the way we can get to the solution.

And manipulating a trig-expression, we change or alter it, convert or modify it, or break it apart, and put the pieces together to get the expression we want so that we can get to the solution.

So algebra connects the problems to the solutions.

In this section therefore, we will get to practice some trig-algebra. And we will get to know some tools called trig-identities, which can help us do trig-algebra, of course.

And in fact, trig-algebra often begins with a trig-identity, which is: $\sin^2 x + \cos^2 x = 1$. What then, is the next?

For instance, dividing both sides by $\cos^2 x$, we get:

$$\frac{\sin^2 x}{\cos^2 x} + \frac{\cos^2 x}{\cos^2 x} = \frac{1}{\cos^2 x} \Rightarrow (\frac{\sin x}{\cos x})^2 + 1 = (\frac{1}{\cos x})^2 \Rightarrow \tan^2 x + 1 = \sec^2 x.$$

And dividing by $\sin^2 x$, both sides of $\sin^2 x + \cos^2 x = 1$, we get:

$$\frac{\sin^2 x}{\sin^2 x} + \frac{\cos^2 x}{\sin^2 x} = \frac{1}{\sin^2 x} \Rightarrow 1 + (\frac{\cos x}{\sin x})^2 = (\frac{1}{\sin x})^2 \Rightarrow \cot^2 x + 1 = \csc^2 x.$$

And also, since we have: $\sin^2 x + \cos^2 x = 1$, we get: $\sin^2 x = 1 - \cos^2 x$. So we get:

$$\tan^2 x \sin^2 x = \tan^2 x (1 - \cos^2 x) = \tan^2 x - \tan^2 x \cos^2 x$$

$$= \tan^2 x - \frac{\sin^2 x}{\cos^2 x} \cos^2 x = \tan^2 x - \sin^2 x.$$

And thus, we get: $\tan^2 x \sin^2 x = \tan^2 x - \sin^2 x.$

So in sum, we have:

$$\sin^2 x + \cos^2 x = 1.$$

$$\tan^2 x + 1 = \sec^2 x. \qquad \cos^2 x + 1 = \csc^2 x.$$

$$\tan^2 x \sin^2 x = \tan^2 x - \sin^2 x.$$

And next, doing some algebra, we can get this: $\dfrac{1 + \tan x}{1 - \tan x} = \dfrac{1 + 2\sin x \cos x}{\cos^2 x - \sin^2 x}.$ How?

Begining with the left hand side, we can get first:

$$\frac{1+\tan x}{1-\tan x} = \frac{(1+\tan x)^2}{(1-\tan x)(1+\tan x)} = \frac{(1+\tan x)^2}{(1-\tan^2 x)}.$$

And we can get: $1-\tan^2 x = 1 - \dfrac{\sin^2 x}{\cos^2 x} = \dfrac{\cos^2 x - \sin^2 x}{\cos^2 x}$, and

$(1+\tan x)^2 = 1 + 2\tan x + \tan^2 x$.   And we know: $1 + \tan^2 x = \sec^2 x$.   So we get:

$$\frac{(1+\tan x)^2}{(1-\tan^2 x)} = \frac{\cos^2 x(\sec^2 x + 2\tan x)}{\cos^2 x - \sin^2 x} = \frac{1 + 2\cos x \sin x}{\cos^2 x - \sin^2 x}, \text{because we have:}$$

$\sec x = \dfrac{1}{\cos x}$, and $\tan x = \dfrac{\sin x}{\cos x}$.

And doing some algebra, we can simpify some expressions. For instance, we can get:

$$\frac{1+\cos x}{\sec x - \tan x} - \frac{1-\cos x}{\sec x + \tan x} = 2(1+\tan x).$$ How come?

Beginning with the left hand side, we get:

$$\frac{1+\cos x}{\sec x - \tan x} - \frac{1-\cos x}{\sec x + \tan x} = \frac{(1+\cos x)(\sec x + \tan x) - (1-\cos x)(\sec - \tan x)}{\sec^2 x - \tan^2 x}.$$

And we can get: $\tan^2 x - \sec^2 x = 1$, since we have: $\tan^2 x + 1 = \sec^2 x$.

And also, we can get: $(1 + \cos x)(\sec x + \tan x) - (1 - \cos x)(\sec x - \tan x)$

$= \sec x + \tan x + \cos x \sec x + \cos x \tan x - (\sec x - \tan x) + \cos x (\sec x - \tan x)$

$= \sec x + \tan x + 1 + \sin x - \sec x + \tan x + 1 - \sin x = 2\tan x + 2 = 2(\tan x + 1)$.

And doing factorization, we can get: $\cos^4 x - \sin^4 x = (\cos^2 x - \sin^2 x)(\cos^2 x + \sin^2 x)$

$= \cos^2 x - \sin^2 x$, simply because we have: $\cos^2 x + \sin^2 x = 1$.

128

What then, about **sin 2x**?

At the end of the example 4 in the set of **Examples in Cosine Functions**, it was mentioned that **sin 2x = 2 cos x sin x**. And we can get it from a trig-identity below:

$$\sin (a + b) = \sin a \cos b + \cos a \sin b.$$

So we get: **sin 2x = sin (x + x) = sin x cos x + cos x sin x = 2 sin x cos x.**

What then, about **sin (a − b)**?

We know: **sin (-b) = -sin b**, and **cos (-b) = cos b**.   So we get:

sin {a + (-b)} = sin (a − b) = sin a cos (-b) + cos a sin (-b) = sin a cos b − cos a sin b.

• Thus, in short, we can set both this way: **sin (a ± b) = sin a cos b ± cos a sin b.**

What then, about **sin 3x**?

Using the identity above, we can get: **sin 3x = sin (2x + x) = sin 2x cos x + cos 2x sin x.**

What then, about **cos 2x**?

We can get it from another trig-identity where: **cos (a + b) = cos a cos b − sin a sin b.**

So we get: **cos 2x = cos x cos x − sin x sin x = cos²x − sin²x ⇒ cos 2x = cos²x − sin²x.**

And we know: $\sin^2 x + \cos^2 x = 1$, so we get: $\cos^2 x = 1 - \sin^2 x$, and $\sin^2 x = 1 - \cos^2 x$.

And thus, we get: $\cos 2x = \cos^2 x - \sin^2 x = 2\cos^2 x - 1 = 1 - 2\sin^2 x$.

And we know: $\sin 2x = 2\cos x \sin x$.   So we get:

$\sin 2x \cos x + \cos 2x \sin x$

$= 2\cos x \sin x \cos x + (1 - 2\sin^2 x)\sin x$

$= 2\sin x \cos^2 x + \sin x - 2\sin^3 x$

$= 2\sin x (1 - \sin^2 x) + \sin x - 2\sin^3 x$

$= 3\sin x - 4\sin^3 x$.

And thus, we get: $\sin 3x = 3\sin x - 4\sin^3 x$.   What then, about $\cos 3x$?

We have: $\cos (a + b) = \cos a \cos b - \sin a \sin b$.

So we get: $\cos 3x = \cos (2x + x) = \cos 2x \cos x - \sin 2x \sin x$.

And we know: $\sin 2x = 2\cos x \sin x$, and $\cos 2x = 2\cos^2 x - 1 = 1 - 2\sin^2 x$.

So we get:

$\cos 2x \cos x - \sin 2x \sin x = (2\cos^2 x - 1)\cos x - 2\cos x \sin x \sin x$

$= 2\cos^3 x - \cos x - 2\cos x \sin^2 x = 2\cos^3 x - \cos x - 2\cos x (1 - \cos^2 x)$

$= 2\cos^3 x - \cos x - 2\cos x + 2\cos^3 x = 4\cos^3 x - 3\cos x$.

And thus, we get: $\cos 3x = 4\cos^3 x - 3\cos x$.

So in sum, we have:

$\sin 3x = 3\sin x - 4\sin^3 x$, and $\cos 3x = 4\cos^3 x - 3\cos x$.

Next, we have a trig-identity where: $\cos(a + b) = \cos a \cos b - \sin a \sin b$.

What then, about $\cos(a - b)$?

We know: $\cos(-b) = \cos b$, and $\sin(-b) = -\sin b$.   So we get:

$\cos\{a + (-b)\} = \cos(a - b) = \cos a \cos(-b) - \sin a \sin(-b) = \cos a \cos b + \sin a \sin b$.

- Thus, in short, we can set both this way: $\cos(a \pm b) = \sin a \cos b \mp \cos a \sin b$.

What then, about $\tan 2x$?

We can get it from another trig-identity where: $\tan(a + b) = \dfrac{\tan a + \tan b}{1 - \tan a \tan b}$.

What then, about $\tan(a - b)$?

We know: $\tan(-b) = -\tan b$. So we get: $\tan(a - b) = \dfrac{\tan a + \tan(-b)}{1 - \tan a \tan(-b)} = \dfrac{\tan a - \tan b}{1 + \tan a \tan b}$.

- Thus, in short, we can set both this way: $\tan(a \pm b) = \dfrac{\tan a \pm \tan b}{1 \mp \tan a \tan b}$.

So we get: $\tan 2x = \dfrac{\tan x + \tan x}{1 - \tan x \tan x} = \dfrac{2\tan x}{1 - \tan^2 x}$.

And in $\cos 2x = 2\cos^2 x - 1 = 1 - 2\sin^2 x$, replacing x with $x/2$, we can get:

$\cos x = 2\cos^2 \frac{x}{2} - 1 = 1 - 2\sin^2 \frac{x}{2}$.

So we can get: $\sin^2 \frac{x}{2} = \dfrac{1 - \cos x}{2}$, and $\cos^2 \frac{x}{2} = \dfrac{1 + \cos x}{2}$.

And we know: $\tan x = \dfrac{\sin x}{\cos x}$.   So we can get: $\tan^2 \frac{x}{2} = \dfrac{\sin^2 \frac{x}{2}}{\cos^2 \frac{x}{2}} = \dfrac{1 - \cos x}{1 + \cos x}$.

What then, about $\tan \frac{x}{2}$?

Since we have: $\tan x = \dfrac{\sin x}{\cos x}$, we can get: $\tan \frac{x}{2} = \dfrac{\sin \frac{x}{2}}{\cos \frac{x}{2}} = \dfrac{\sin \frac{x}{2} \cos \frac{x}{2}}{\cos^2 \frac{x}{2}}$.

And we have: $\sin 2x = 2\sin x \cos x$. So we get: $\sin x = 2\sin \frac{x}{2} \cos \frac{x}{2}$.

And also, we have: $\cos^2 \frac{x}{2} = \dfrac{1 + \cos x}{2}$.   So we get: $2\cos^2 (x/2) = 1 + \cos x$.

And thus, we get: $\tan \frac{x}{2} = \dfrac{\sin \frac{x}{2} \cos \frac{x}{2}}{\cos^2 \frac{x}{2}} = \dfrac{2\sin \frac{x}{2} \cos \frac{x}{2}}{2\cos^2 \frac{x}{2}} = \dfrac{\sin x}{1 + \cos x} \Rightarrow \tan \frac{x}{2} = \dfrac{\sin x}{1 + \cos x}$.

And from $\sin (a \pm b)$ and $\cos (a \pm b)$, we can get some useful expressions as follows:

To begin with, we have: $\sin (a \pm b) = \sin a \cos b \pm \cos a \sin b$.   So we get:

$\sin (a + b) + \sin (a - b) = \sin a \cos b + \cos a \sin b + \sin a \cos b - \cos a \sin b$

$= 2\sin a \cos b$.   And thus, we get: $\sin a \cos b = \frac{1}{2} \{\sin (a + b) + \sin (a - b)\}$.

Next, $\sin(a+b) - \sin(a-b) = \sin a \cos b + \cos a \sin b - \sin a \cos b + \cos a \sin b$

$= 2\cos a \sin b$. So we get: $\cos a \sin b = \frac{1}{2}\{\sin(a+b) - \sin(a-b)\}$.

Next, $\cos(a+b) + \cos(a-b) = \cos a \cos b - \sin a \sin b + \cos a \cos b + \sin a \sin b$

$= 2\cos a \cos b$.   So we get: $\cos a \cos b = \frac{1}{2}\{\cos(a+b) + \cos(a-b)\}$.

And next, $\cos(a+b) - \cos(a-b) = \cos a \cos b - \sin a \sin b - \cos a \cos b - \sin a \sin b$

$= -2\sin a \sin b$.   So we get: $\sin a \sin b = -\frac{1}{2}\{\cos(a+b) + \cos(a-b)\}$.

And also, we can put the expressions above the way below, too:

To begin with, we have: $\sin(a+b) + \sin(a-b) = 2\sin a \cos b$.

And setting $x = a + b$, and $y = a - b$, we can get: $a = \frac{x+y}{2}$, and $b = \frac{x-y}{2}$.

So we get: $\sin(a+b) + \sin(a-b) = \sin x + \sin y$, and $2\sin a \cos b = 2\sin\frac{x+y}{2}\cos\frac{x-y}{2}$.

And thus, we get: $\sin x + \sin y = 2\sin\frac{x+y}{2}\cos\frac{x-y}{2}$.

Next, we have: $\sin(a+b) - \sin(a-b) = 2\cos a \sin b$, $a = \frac{x+y}{2}$, and $b = \frac{x-y}{2}$.
So we get: $\sin x - \sin y = 2\cos\frac{x+y}{2}\sin\frac{x-y}{2}$.

Next, we have: $\cos(a+b) + \cos(a-b) = 2\cos a \cos b$, $a = \frac{x+y}{2}$, and $b = \frac{x-y}{2}$.
So we get: $\cos x + \cos y = 2\cos\frac{x+y}{2}\cos\frac{x-y}{2}$.

And next, we have: $\cos(a+b) - \cos(a-b) = -2\sin a \sin b$, $a = \frac{x+y}{2}$, and $b = \frac{x-y}{2}$.

So we get: $\cos x - \cos y = -2\sin\frac{x+y}{2}\sin\frac{x-y}{2}$.

And using the distance formula, together with **sin** $(a \pm b)$ and **cos** $(a \pm b)$, we can get a couple of useful expressions. And they are as follows:

• Assuming first, in the **x-y** plane, **p** is the angle between the **x**-axis to the right of the origin and a line segment connecting the origin and a point (u, v) in the first quadrant, we get:

$$u \sin x + v \cos x = \sqrt{u^2 + v^2}\, \sin(x+p), \text{ where } \cos p = \frac{u}{\sqrt{u^2+v^2}}, \text{ and } \sin p = \frac{v}{\sqrt{u^2+v^2}}.$$

• And assuming next, in the **x-y** plane, **q** is the angle between the **x**-axis to the right of the origin and a line segment connecting the origin and a point (s, t) in the first quadrant, we get:

$$s \cos x + t \sin x = \sqrt{s^2 + t^2}\, \cos(x-q), \text{ where } \cos q = \frac{s}{\sqrt{s^2+t^2}}, \text{ and } \sin q = \frac{t}{\sqrt{s^2+t^2}}.$$

And let's see now how it is the case.

To begin with, putting the point (u, v) in the **x-y** plane, we get:

Fig. 0

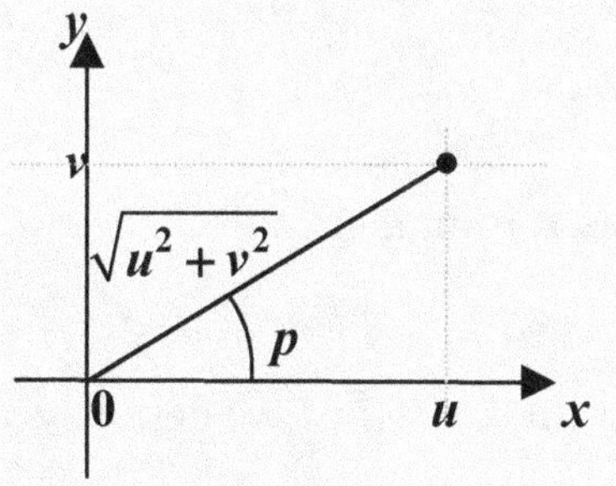

So we get: $\cos p = \frac{u}{\sqrt{u^2+v^2}}$, and $\sin p = \frac{v}{\sqrt{u^2+v^2}}$.

And next, we can set: $u \sin x + v \cos x = \sqrt{u^2 + v^2}\, (\frac{u}{\sqrt{u^2+v^2}} \sin x + \frac{v}{\sqrt{u^2+v^2}} \cos x)$.

So we get: $u \sin x + v \cos x = \sqrt{u^2 + v^2}\, (\cos p \sin x + \sin p \cos x)$.

And we know: $\sin (x + p) = \sin p \cos x + \cos p \sin x$.   Thus, we get:

$u \sin x + v \cos x = \sqrt{u^2 + v^2}\, \sin (x + p)$, where $\cos p = \frac{u}{\sqrt{u^2+v^2}}$, and $\sin p = \frac{v}{\sqrt{u^2+v^2}}$.

And next, putting the point (s, t) in the x-y plane, we get:

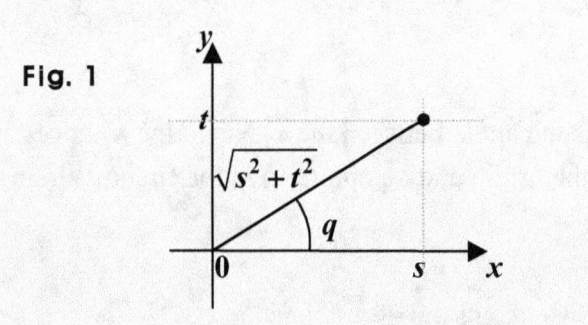

Fig. 1

So we get: $\cos q = \frac{s}{\sqrt{s^2+t^2}}$, and $\sin q = \frac{t}{\sqrt{s^2+t^2}}$.

And we can set: $s \cos x + t \sin x = \sqrt{s^2 + t^2}\, (\frac{s}{\sqrt{s^2+t^2}} \cos x + \frac{t}{\sqrt{s^2+t^2}} \sin x)$.

So we get: $s \cos x + t \sin x = \sqrt{s^2 + t^2}\, (\cos q \cos x + \sin q \sin x)$.

And we know: $\cos (x - q) = \cos x \cos q + \sin x \sin q$.   Thus, we get:

$s \cos x + t \sin x = \sqrt{s^2 + t^2}\, \cos (x - q)$, where $\cos q = \frac{s}{\sqrt{s^2+t^2}}$, and $\sin q = \frac{t}{\sqrt{s^2+t^2}}$.

And of course, using **sin ($a \pm b$)**, **cos ($a \pm b$)**, and other trig-identities, along with factorization, we can change many other trig-expressions, too. For instance:

sin ($a + b$) sin ($a - b$) = (sin a cos b + cos a sin b)(sin a cos b – cos a sin b)

$= \sin^2 a \cos^2 b - \cos^2 a \sin^2 b = \sin^2 a \,(1 - \sin^2 b) - (1 - \sin^2 a)\sin^2 b$

$= \sin^2 a - \sin^2 a \sin^2 b - \sin^2 b + \sin^2 a \sin^2 b = \sin^2 a - \sin^2 b$

$= (1 - \cos^2 a) - (1 - \cos^2 b) = \cos^2 b - \cos^2 a.$

So we can get: **sin ($a + b$) sin ($a - b$) = $\sin^2 a - \sin^2 b = \cos^2 b - \cos^2 a$.**

And let's next, try using the equality above:

We are going to put $\dfrac{\sin^2 a - \sin^2 b}{\sin^2 (a + b)}$ in terms of **tan a** and **tan b**.

Then, we get: $\dfrac{\sin^2 a - \sin^2 b}{\sin^2(a+b)} = \dfrac{\sin(a+b)\sin(a-b)}{\sin^2(a+b)} = \dfrac{\sin(a-b)}{\sin(a+b)} = \dfrac{\sin a \cos b - \cos a \sin b}{\sin a \cos b + \cos a \sin b}$

$= \dfrac{(\sin a \cos b - \cos a \sin b)\frac{1}{\cos a \cos b}}{(\sin a \cos b + \cos a \sin b)\frac{1}{\cos a \cos b}} = \dfrac{\frac{\sin a}{\cos a} - \frac{\sin b}{\cos b}}{\frac{\sin a}{\cos a} + \frac{\sin b}{\cos b}} = \dfrac{\tan a - \tan b}{\tan a + \tan b}.$

And let's next, try changing **cos ($a + b$) cos ($a - b$)** the way we did for the sines above.

cos ($a + b$) cos ($a - b$) = (cos a cos b – sin a sin b)(cos a cos b + sin a sin b)

$= \cos^2 a \cos^2 b - \sin^2 a \sin^2 b = \cos^2 a \,(1 - \sin^2 b) - (1 - \cos^2 a)\sin^2 b$

$= \cos^2 a - \cos^2 a \sin^2 b - \sin^2 b + \cos^2 a \sin^2 b = \cos^2 a - \sin^2 b$

$= (1 - \sin^2 a) - (1 - \cos^2 b) = \cos^2 b - \sin^2 a.$

So we can get: **cos ($a + b$) cos ($a - b$) = $\cos^2 a - \sin^2 b = \cos^2 b - \sin^2 a$.**

And let's next, for another instance, see what we can get simplifying a trig-expression as follows: $\cos a \sin (b - c) + \cos b \sin (c - a) + \cos c \sin (a - b)$.

$$\cos a \sin (b - c) = \cos a (\sin b \cos c - \cos b \sin c) = \cos a \cos c \sin b - \cos a \cos b \sin c$$

$$\cos b \sin (c - a) = \cos b (\sin c \cos a - \cos c \sin a) = \cos a \cos b \sin c - \cos b \cos c \sin a$$

$$\cos c \sin (a - b) = \cos c (\sin a \cos b - \cos a \sin b) = \cos b \cos c \sin a - \cos a \cos c \sin b$$

So we can get: $\cos a \sin (b - c) + \cos b \sin (c - a) + \cos c \sin (a - b) = 0$.

And let's next, for another instance, see what we can get simplifying a trig-expression as follows: $\cos^2 (x + \pi/4) + \cos^2 (x - \pi/4)$.

To begin with, we have: $\cos (x + \pi/4) = \cos x \cos \pi/4 - \sin x \sin \pi/4 = \frac{\sqrt{2}}{2} (\cos x - \sin x)$.

And we have: $\cos (x - \pi/4) = \cos x \cos \pi/4 + \sin x \sin \pi/4 = \frac{\sqrt{2}}{2} (\cos x + \sin x)$.

So we get:

$\cos^2 (x + \pi/4) = \frac{1}{2} (\cos^2 x - 2\cos x \sin x + \sin^2 x)$.

$\cos^2 (x - \pi/4) = \frac{1}{2} (\cos^2 x + 2\cos x \sin x + \sin^2 x)$.

Thus, we get: $\cos^2 (x + \pi/4) + \cos^2 (x - \pi/4) = \cos^2 x + \sin^2 x = 1$.

What then, about $3 \cos^2 (x + \pi/3) + \cos^2 (x - \pi/3)$?

We have: $\cos(x + \pi/3) = \cos x \cos \pi/3 - \sin x \sin \pi/3 = \frac{1}{2}(\cos x - \sqrt{3} \sin x)$.

And we have: $\cos(x - \pi/3) = \cos x \cos \pi/3 + \sin x \sin \pi/3 = \frac{1}{2}(\cos x + \sqrt{3} \sin x)$.

So we get:

$\cos^2(x + \pi/3) = \frac{1}{4}(\cos^2 x - 2\sqrt{3} \cos x \sin x + 3\sin^2 x)$.

$\cos^2(x - \pi/3) = \frac{1}{4}(\cos^2 x + 2\sqrt{3} \cos x \sin x + 3\sin^2 x)$.

Thus, we get: $3\cos^2(x + \pi/4) + \cos^2(x - \pi/4) = \frac{3}{2}(\cos^2 x + \sin^2 x) = \frac{3}{2} \cdot 1 = \frac{3}{2}$.

Examples in Trig–Algebra

Note: $(\sin x)(\cos x) = \sin x \cos x$, and $(\sin x)^n = \sin^n x$, so for instance, $(\sin x)^2 = \sin^2 x$.

0.   Assuming $\sin x + \cos x = 1/2$, find the values of the expressions below:

$\sin x - \cos x$, $\sin^3 x + \cos^3 x$, $\sin^6 x + \cos^6 x$, $\tan x + \cot x$, and $\tan^3 x + \cot^3 x$.

1.0.   Assuming $\sin x + \cos x = \sin x \cos x = a$, find a and the value of $\tan x + \cot x$.

1.1.   Assuming $\sin x + \cos x = b$, and $\sin x \cos x = (b^2 - 2)/4$, find b and the value of $\tan x + \cot x$.

1.2.   Assuming $\sin x + \cos x = 0,$ find the sum and the product of the two in each pair as follows: $\sin^2 x$ and $\cos^2 x$, $\sin^3 x$ and $\cos^3 x$, and $\tan x$ and $\cot x$.

2.0.   Assuming $\sin x = 3/5$ and $\pi/2 < x < \pi$, find $\cos x$ and $\tan x$.

2.1.   Assuming $\tan x = -5/12$, find $\sin x$ and $\cos x$.

2.2.   Assuming $2\cos x - \sin x = 1$ and $0 < x < \pi/2$, find $\sin x$ and $\cos x$.

2.3.   Assuming $6\sin^2 x + \sin x \cos x - 2\cos^2 x = 0$ and $\pi/2 < x < \pi$, find $\sin x + \cos x$ and $\tan x$.

3. Assuming *a*, *b*, *u*, and *v* are constants, find in each case below, the connective expression between the constants:

3.0. $a \sin x - b \cos x = 1$, and $b \sin x + a \cos x = 1 + b \cos x$.

3.1. $u = a \sin^3 x$, and $v = b \cos^3 x$ where $a \neq 0$ and $b \neq 0$.

3.2. $a \sin x + \cos x = 1$, and $b \sin x - \cos x = 1$.

3.3. $a \sec x = 1 + \tan x$, and $b \sec x = 1 - \tan x$.

4. Find the angle between two lines $y = ax + b$ and $y = cx + d$.

Suggestions or Solutions
To the **Problem** in the Example **0**

Assuming sin x + cos x = 1/2, find the values of the expressions below:
sin x – cos x, sin^3 x + cos^3 x, sin^6 x + cos^6 x, tan x + cot x, and tan^3 x + cot^3 x.

Doing algebra, we often do polynomial factorizations. The same is true for trig-algebra, too. And quite often, the trig-identity where **sin^2 x + cos^2 x = 1** is useful.

So for instance, taking the square of the first expression, we get:

(sin x – cos x)^2 = sin^2 x + cos^2 x – 2 sin x cos x = 1 – 2 sin x cos x.

And thus, knowing the value of **sin x cos x**, we can get the value of **sin x – cos x.**

To begin with, for simplicity, setting: s = **sin x**, and c = **cos x**, we have: $s + c$ = **1/2**.

So we can get: **$(s + c)^2 = s^2 + c^2 + 2sc = 1 + 2sc = 1/4$.**

Thus, we get: **$sc = (1/4 – 1)/2 = -3/8$.**

So next, beginning with: **sin x – cos x**, we get:

$(s – c)^2 = s^2 + c^2 – 2sc = 1 – 2(-3/8) = 1 + 6/8 = 14/8 = 7/4 \Rightarrow (s – c)^2 = 7/4$.

Thus, we get: **sin x – cos x = $\pm \frac{\sqrt{7}}{2}$.**

And using sc = **-3/8**, along with factorization, we get:

$s^3 + c^3 = (s + c)^3 – 3sc(s + c) = (1/2)^3 – 3(-3/8)(1/2) = 1/8 + 9/16 = 11/16$.

$s^6 + c^6 = (s^3 + c^3)^2 – 2s^3c^3 = (11/16)^2 – 2(-3/8)^3 = 11^2/2^8 + 27/2^8 = (121+27)/2^8$

$= 148/2^8 = 2 \cdot 74/2^8 = 4 \cdot 37/2^8 = 37/2^6 = 37/64$.

And we can get the same the way below, too:

$$s^6 + c^6 = (s^2 + c^2)^3 - 3s^2c^2(s^2 + c^2) = 1 - 3(-3/8)^2 \cdot 1 = (64 - 27)/64 = 37/64.$$

Next, $\tan x + \cot x = s/c + c/s = (s^2 + c^2)/(sc) = 1/(-3/8) = -8/3$.

And next:

$$\tan^3 x + \cot^3 x = (\tan x + \cot x)^3 - 3 \tan x \cot x (\tan x + \cot x)$$

$$= (-8/3)^3 - 3 \cdot 1 \cdot (-8/3) = -2^9/3^3 + 2^3 = (3^3 2^3 - 2^9)/3^3 = (27 \cdot 8 - 16 \cdot 32)/27$$

$$= (240 - 24 - 320 - 180 - 12)/27 = (-200 + 40 - 124 - 12)/27 = -296/27.$$

Suggestions or Solutions
To the **Problems** in the Example 1

1.0. Assuming $\sin x + \cos x = \sin x \cos x = a$, find a and the value of $\tan x + \cot x$.

Assuming first, for simplicity, $s = \sin x$, and $c = \cos x$, we have: $s + c = a$, and $sc = a$.

Then, we get: $(s + c)^2 = s^2 + c^2 + 2sc \Rightarrow a^2 = 1 + 2a \Rightarrow a^2 - 2a - 1 = 0$.

So next, using the quadratic formula, we get: $a = 1 \pm \sqrt{1+1} = 1 \pm \sqrt{2}$.

So do we have: $a = 1 + \sqrt{2}$ or $1 - \sqrt{2}$?

We have this: $-1 \leq \sin\theta \leq 1$, and $-1 \leq \cos\theta \leq 1$.

So we get: $-1 \leq a \leq 1$, because $sc = a$. And thus, we get: $a = 1 - \sqrt{2}$.

So next, we get: $\tan x + \cot x = s/c + c/s = (s^2 + c^2)/(sc) = 1/a$

$$= \frac{1}{1-\sqrt{2}} = \frac{1+\sqrt{2}}{(1-\sqrt{2})(1+\sqrt{2})} = \frac{1+\sqrt{2}}{1-2} = -(1+\sqrt{2}).$$

1.1.   assuming $\sin x + \cos x = b$, and $\sin x \cos x = (b^2 - 2)/4$, find b and the value of $\tan x + \cot x$.

Setting first, for simplicity, $s = \sin x$, and $c = \cos x$, we get: $s + c = b$, and $sc = (b^2 - 2)/4$.

So we get: $(s + c)^2 = b^2 = s^2 + c^2 + 2sc = 1 + (b^2 - 2)/2 \Rightarrow 2b^2 = 2 + b^2 - 2 \Rightarrow b = 0$.

Thus, next, we get: $\tan x + \cot x = s/c + c/s = (s^2 + c^2)/(sc) = 1/(sc) = 4/(b^2 - 2) = -2$.

1.2.   Assuming $\sin x + \cos x = 0$, find the sum and the product of the two in each pair as follows: $(\sin^2 x$ and $\cos^2 x)$, $(\sin^3 x$ and $\cos^3 x)$, and $(\tan x$ and $\cot x)$.

Setting first, for simplicity, $s = \sin x$, and $c = \cos x$, we have: $s + c = 0$, and, we can get:

$(s + c)^2 = s^2 + c^2 + 2sc \Rightarrow 0 = 1 + 2sc$, since we have: $s^2 + c^2 = 1$, which is a trig-identity, and we are given: $s + c = 0$.

Thus, we get: $sc = -1/2$.

So beginning with $\sin^2 x$ and $\cos^2 x$, we get: $s^2 + c^2 = 1$, and $s^2 c^2 = 1/4$.

Next, moving on to $\sin^3 x$ and $\cos^3 x$, we get: $s^3 + c^3 = (s + c)^3 - 3sc(s + c) = 0$, since we have: $s + c = 0$.

And next, $\tan x + \cot x = s/c + c/s = (s^2 + c^2)/(sc) = -2$, and $\tan x \cot x = (s/c)(c/s) = 1$.

Suggestions or Solutions
To the **Problems** in the Example **2**

2.0.   Assuming sin x = 3/5 and $\pi/2 < x < \pi$, find cos x and tan x.

We know that the sine is the ratio of the opposite to the hypotenuse in a right triangle.

In short, **sin x** is: the opposite over the hypotenuse.

And know that the opposite faces the angle x, because x is the governing angle.

So from **sin x = 3/5**, we can see that the ratio of the opposite to the hypotenuse is 3:5.

And thus, assuming **O** is the opposite, **H** is the hypotenuse, and **A** is the adjacent, we get: **O/H = 3/5**.

So assuming **O = 3a** where a is a constant > 0, we get: **H = 5a**.

What then, about the adjacent **A**?

We know that **O, H**, and **A** are three sides in a right triangle.

So we can get A using the distance formula, called Pythagorean theorems, too.
And thus, using the formula, we get:

$$H^2 = O^2 + A^2 \Rightarrow (5a)^2 = (3a)^2 + A^2 \Rightarrow A^2 = (25-9)a = 16a \Rightarrow A = 4a.$$

And we know that the cosine is: the adjacent over the hypotenuse, and that the tangent is: the opposite over the adjacent.

So we get: **cos x = A/H = 4a/5a = 4/5**, and **tan x = O/A = 3a/4a = 3/4**.

We have however, $\pi/2 < x < \pi$. So we get: **-1 < cos x < 0**, and **tan x < 0**.

Thus, we get: **cos x = -4/5**, and **tan x = -3/4**.

2.1. Assuming tan x = -5/12, find sin x and cos x.

From **tan** x = **-5/12**, we can see that the opposite over the adjacent = -5/12, which is negative. So the right triangle is not ordinary but transcendental.

And taking the absolute value of the **tan** above, we get: $|\text{tan } x|$ = |-5/12| = 5/12,

And thus, assuming O is the opposite, H is the hypotenuse, and A is the adjacent, we get: $|O/A|$ = **5/12**.

So assuming O = **5a** where a is a constant > 0, we get: A = **12a**.

And we can get H using the distance formula. So using the formula, we get:

$$H^2 = O^2 + A^2 = (5a)^2 + (12a)^2 \Rightarrow H^2 = (25 + 144)a = 169a \Rightarrow H = 13a.$$

And we know that the sine is: the opposite over the hypotenuse, and that the cosine is: the adjacent over the hypotenuse.

And also, we know that the x in **tan** x is the governing angle, and that the ray of length 1 turning about the origin in the x-y plane makes the governing angle x.

And the sign of the trig-ratio depends on what quadrant the ray is in. For instance, if the ray is in the second quadrant, the sign of the sine is positive, but the signs of the cosine and the tangent are negative.

So we can set for now:

$|\text{sin } x|$ = $|O/H|$ = 5a/13a = 5/13, and $|\text{cos } x|$= $|O/A|$ = 12a/13a = 12/13.

And we know that if **tan** x **< 0**, the ray is in the second or the fourth quadrant.

And we know if the ray is in the second quadrant, the sine is positive, but the cosine is negative. And if the ray is in the fourth quadrant, the sine is negative but the cosine is positive. So given **tan** x = **-5/12**, we can get:

sin x = **5/13** and **cos** x = **-12/13**, or **sin** x = **-5/13** and **cos** x = **12/13**.

2.2.   Assuming $2\cos x - \sin x = 1$ and $0 < x < \pi/2$, find $\sin x$ and $\cos x$.

Using the equation given, we can get an equation for **cos x**. Solving the equation then, we can get the value of **cos x**.   How then, can we get the equation?

We have a trig-identity: $\sin^2 x + \cos^2 x = 1$. And we have this, too: $2\cos x - \sin x = 1$.

So we can put the identity in terms of **cos x**, and then, we get the equation for **cos x**.

To begin with, we get: $2\cos x - \sin x = 1 \Rightarrow \sin x = 2\cos x - 1$.

So we get: $\sin^2 x + \cos^2 x = (2\cos x - 1)^2 + \cos^2 x = 4\cos^2 x - 4\cos x + 1 + \cos^2 x$

$= 5\cos^2 x - 4\cos x + 1 = 1 \Rightarrow \cos x \,(5\cos x - 4) = 0 \Rightarrow \cos x = 0$ or $4/5$.

And we have this, too: $0 < x < \pi/2$.   So we get: $\cos x = 4/5$.

What then, about **sin x**?

We have: $\sin x = 2\cos x - 1$.   So we get: $\sin x = 2(4/5) - 1 = 3/5$.

What then, about **tan x**?

We know: **tan x = sin x / cos x**.   So we get: $\tan x = (3/5)/(4/5) = 3/4$.

2.3. **Assuming** $6\sin^2 x + \sin x \cos x - 2\cos^2 x = 0$ **and** $\pi/2 < x < \pi$**, find** $\sin x + \cos x$ **and** $\tan x$**.**

Examining the equation give, we can modify it into an equation for **tan x**.

Solving the equation then, we can get the value of **tan x**.

So to begin with, dividing by $\cos^2 x$ the equation $6\sin^2 x + \sin x \cos x - 2\cos^2 x = 0$, we get: $6\tan^2 x + \tan x - 2 = 0$.

And next, factorizing it, we can get: $(3\tan x + 2)(2\tan x - 1) = 0$.

So we get: **tan x = -2/3 or 1/2**.

And we have this too: $\pi/2 < x < \pi$. That is, the ray is in the second quadrant.

So we can see that **tan x = -2/3**. What then, about **sin x + cos x**?

We know that the tangent is: the opposite over the hypotenuse:

And we can get this: $|\tan x| = |-2/3| = 2/3$.

So we get: **|the opposite|** = 2, and **|the adjacent|** = 3.

And by the distance formula, we get: **the hypotenuse** $= \sqrt{2^2 + 3^2} = \sqrt{13}$.

So we get: $|\sin x| = \frac{2}{\sqrt{13}}$, and $|\cos x| = \frac{3}{\sqrt{13}}$. And we have: $\pi/2 < x < \pi$.

Thus, we get: $\sin x + \cos x = \frac{2}{\sqrt{13}} - \frac{3}{\sqrt{13}} = -\frac{1}{\sqrt{13}} = -\frac{\sqrt{13}}{13}$.

By the way, solving a trig-equation $\sin x = 1$ with no more specification, we don't want to put the solution this way: $x = \pi/2$.

That's because it's not the only solution.

We can get: $\sin x = 1$ when not only $x = \pi/2$ but $x = 2\pi + \pi/2$, $-\pi - \pi/2$, $-\pi/2$, etc., too.

And in fact, we can get: $\sin x = 1$ when $x = n\pi + (-1)^n(\pi/2)$ for n integer.

So for instance:

$\sin x = 1/2 \Rightarrow x = n\pi + (-1)^n(\pi/6)$ for n integer.

$\sin 2x = 1/2 \Rightarrow 2x = n\pi + (-1)^n(\pi/6)$ for n integer $\Rightarrow x = n\pi/2 + (-1)^n(\pi/12)$ for n integer.

$\sin x = -1/2 \Rightarrow x = n\pi + (-1)^n(-\pi/6)$ for n integer $\Rightarrow x = n\pi - (-1)^n(\pi/6)$ for n integer.

$\cos x = 1/2 \Rightarrow x = 2n\pi \pm (\pi/3)$ for n integer.

$\tan x = \frac{\sqrt{2}}{2} \Rightarrow x = n\pi + (\pi/4)$ for n integer.

That is to say that:

$\sin x = \sin a \Rightarrow x = n\pi + (-1)^n a$ for n integer.

$\cos x = \cos b \Rightarrow x = 2n\pi \pm b$ for n integer.

$\tan x = \tan c \Rightarrow x = n\pi + c$ for n integer.

Suggestions or Solutions
To the Problems in the Example 3

Assuming a, b, u, and v are constants, find in each case below, the connective expression between the constants:

3.0. $a \sin x - b \cos x = 1$, and $b \sin x + a \cos x = 1 + b \cos x$.

3.1. $u = a \sin^3 x$, and $v = b \cos^3 x$ where $a \neq 0$ and $b \neq 0$.

3.2. $a \sin x + \cos x = 1$, and $b \sin x - \cos x = 1$.

3.3. $a \sec x = 1 + \tan x$, and $b \sec x = 1 - \tan x$.

3.0. Assuming $s = \sin x$, and $c = \cos x$, we get: $as - bc = 1$, and $bs + ac = 1 + bc$.

So we get: $as - bc = 1$ --- **(1)**, and $bs + (a - b)c = 1$ --- **(2)**.

Let's next, remove c first.

Multiplying (1) by $(a - b)$, we get: $a(a - b)s - b(a - b)\, c = 1(a - b) = a - b$.

Next, multiplying (2) by b, we get: $b^2 s + b(a - b)c = b$.

So next, adding together the two equations above, we get:

$$(a - b)as + b^2 s = a - b + b \Rightarrow (a^2 - ab + b^2)s = a.$$

And we can put the processes above the way below, too:
$$(1) \cdot (a - b) + (2) \cdot b \Rightarrow a(a - b)s + b^2 s = (a - b) + b = a \Rightarrow (a^2 - ab + b^2)s = a.$$

So next, removing s from the two equations (1) and (2), we can do it the way below:

$$(1) \cdot b - (2) \cdot a = -b^2 c - a(a - b)c = b - a \Rightarrow (-b^2 - a^2 + ab)c = b - a$$

$$\Rightarrow (a^2 - ab + b^2)c = a - b. \text{ What then, is the next?}$$

We have: $s^2 + c^2 = 1$, and the two equations as follows:

$(a^2 - ab + b^2)s = a$, and $(a^2 - ab + b^2)c = a - b$.

So we can get the connective expression we want the way below:

Solving first, for s the equation $(a^2 - ab + b^2)s = a$, and also, solving for c the equation $(a^2 - ab + b^2)c = a - b$.

And next, putting into the identity $s^2 + c^2 = 1$, the expression for s and the expression for c, we can get the connective expression we want.

We want to make sure first though, $a^2 - ab + b^2$ is not 0. So checking it, we get:

$a^2 - ab + b^2 = a^2 - ab + b^2/4 - b^2/4 + b^2 = (a - b/2)^2 + 3b^2/4 > 0$.

That's because a and b cannot be 0 at the same time due to the equation (1) above.

And thus, we can get:

$(a^2 - ab + b^2)s = a \Rightarrow s = a/(a^2 - ab + b^2)$.

$(a^2 - ab + b^2)c = a - b \Rightarrow c = (a - b)/(a^2 - ab + b^2)$.

So next, putting the s and the c above into the identity $s^2 + c^2 = 1$, we get:

$\{ a/(a^2 - ab + b^2)\}^2 + \{(a - b)/(a^2 - ab + b^2)\}^2 = \{a^2 + (a - b)^2\}/(a^2 - ab + b^2)^2 = 1$, which is the connective expression.

And of course, we can put the connective expression the way below, too:

$(2a^2 - 2ab + b^2) = (a^2 - ab + b^2)^2$.

3.1. $u = a \sin^3 x$, **and** $v = b \cos^3 x$ **where** $a \neq 0$ **and** $b \neq 0$.

To begin with, we can get:

$u = a \sin^3 x \Rightarrow \sin x = (u/a)^{1/3} = \sqrt[3]{\frac{u}{a}}$, and $v = b \cos^3 x \Rightarrow \cos x = (v/b)^{1/3} = \sqrt[3]{\frac{v}{b}}$.

So next, we get: $\sin^2 x + \cos^2 x = (u/a)^{2/3} + (v/b)^{2/3} = 1$.

And thus, the connective expression is: $(u/a)^{2/3} + (v/b)^{2/3} = 1$.

3.2. $a \sin x + \cos x = 1$, **and** $b \sin x - \cos x = 1$.

Assuming first, $s = \sin x$, and $c = \cos x$, we get: $as + c = 1$ --- **(1)**, and $bs - c = 1$ --- **(2)**.

So next, we can get:

(1) + (2) $\Rightarrow (a + b)s = 2 \Rightarrow s = 2/(a + b)$.

(1) $\cdot b -$ **(2)** $\cdot a = bc + ac = b - a \Rightarrow c = (b - a)/(a + b)$.

Thus, next, we can get:

$s^2 + c^2 = 1 \Rightarrow 4/(a + b)^2 + (b - a)^2/(a + b)^2 = \{4 + (b - a)^2\}/(a + b)^2 = 1$

$\Rightarrow 4 + b^2 - 2ab + a^2 = a^2 + 2ab + b^2 \Rightarrow 4ab = 4 \Rightarrow ab = 1$, which is the connective expression.

3.3. *a* sec *x* = 1 + tan *x*, and *b* sec *x* = 1 – tan *x*.

Assuming first, *s* = **sin** *x*, and *c* = **cos** *x*, we get:

$a/c = 1 + s/c \Rightarrow a = c + s$ --- **(1)**, and $b/c = 1 - s/c \Rightarrow b = c - s$ --- **(2)**.

So we get:

(1) + **(2)** $\Rightarrow a + b = 2c \Rightarrow c = (a + b)/2$, and **(1)** – **(2)** $\Rightarrow a - b = 2s \Rightarrow s = (a - b)/2$.

Thus, we get: $s^2 + c^2 = 1 \Rightarrow (a - b)^2/4 + (a + b)^2/4 = (a^2 - 2ab + b^2 + a^2 + 2ab + b^2)/4$

$= 2(a^2 + b^2)/4 = (a^2 + b^2)/2 = 1 \Rightarrow a^2 + b^2 = 2$, which is the connective expression.

Suggestions or Solutions
To the Problem in the Example 4

Find the angle between two lines $y = ax + b$ **and** $y = cx + d$**.**

Suppose first, **tan** $u = a$, **tan** $v = c$, and $0 < a < c$.

Then, putting the two lines in the **x-y** plane, we can put them the way below:

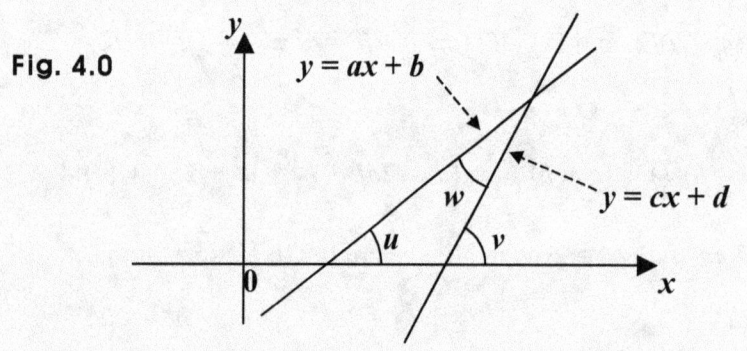

Fig. 4.0

And we know that $v = u + w$, because $u + v + w = \pi$.

That is to say that we have: $w = v - u$.

Then, we get: **tan** $w =$ **tan** $(v - u)$.

And we have: $\tan (v \pm u) = \dfrac{\tan v \pm \tan u}{1 \mp \tan v \tan u}$.

So we get: $\tan w = \tan (v - u) = \dfrac{\tan v - \tan u}{1 + \tan v \tan u} = \dfrac{c - a}{1 + ca}$.

And thus, we get: $w = \tan^{-1} \dfrac{c - a}{1 + ca}$.

Suppose next, **tan** $u = a$, **tan** $v = c$, $a > 0$, and $c < 0$.

Then, putting the two lines in the **x-y** plane, we can put them the way below:

Fig. 4.1

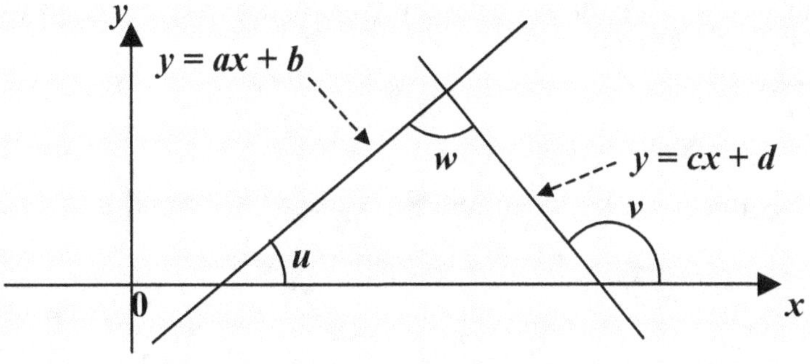

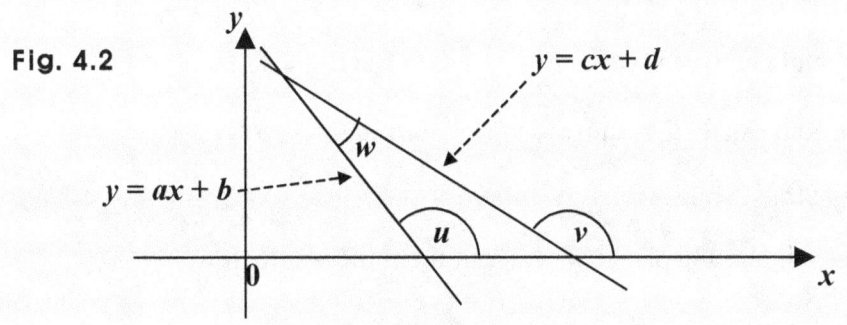

Then again, since $v = u + w$, we get: $w = v - u$, too. So we get: **tan** w = **tan** $(v - u)$.

Thus, we get: **tan** w = **tan** $(v - u) = \dfrac{\tan v - \tan u}{1 + \tan v \tan u} = \dfrac{c - a}{1 + ca}$.

So we get: $w = \tan^{-1} \dfrac{c - a}{1 + ca}$.

Suppose next, **tan** $u = a$, **tan** $v = c$, and $a < c < 0$.

Then, putting the two lines in the **x-y** plane, we can put them the way below:

Fig. 4.2

Then again, since $v = u + w$, we get: $w = v - u$, too. So we get: **tan** w = **tan** $(v - u)$.

Thus, we get: **tan** w = **tan** $(v - u) = \dfrac{\tan v - \tan u}{1 + \tan v \tan u} = \dfrac{c - a}{1 + ca}$.

So we get: $w = \tan^{-1} \dfrac{c - a}{1 + ca}$.

So we can say that assuming w is the angle between two lines $y = ax + b$ and $y = cx + d$, we get: $w = \tan^{-1} \dfrac{c - a}{1 + ca}$.

Suppose for instance, w is the angle between two lines $y = \frac{\sqrt{3}}{3}x - 1$ and $y = \sqrt{3}x - 9$.

Then, $\tan \pi/6 = \frac{\sqrt{3}}{3} = a$, and $\tan \pi/3 = \sqrt{3} = c$. That is, $u = \pi/6$, and $v = \pi/3$. So $w = \pi/6$.

And putting the two lines in the x-y plane, we can put them the way below:

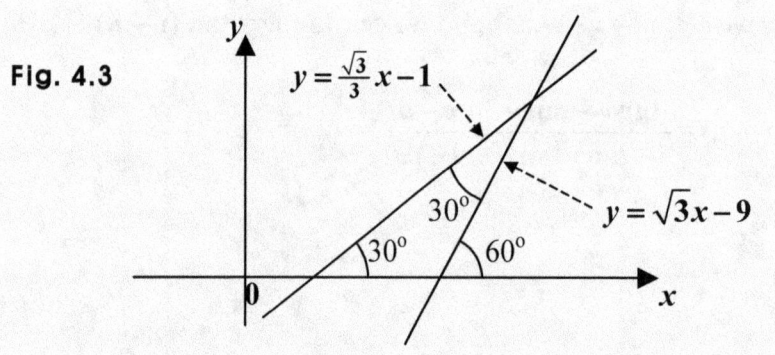

Fig. 4.3

Then, we get: $\dfrac{c - a}{1 + ca} = \dfrac{\sqrt{3} - \frac{\sqrt{3}}{3}}{1 + \sqrt{3} \cdot \frac{\sqrt{3}}{3}} = \dfrac{\frac{3\sqrt{3} - \sqrt{3}}{3}}{2} = \dfrac{2\sqrt{3}}{6} = \dfrac{\sqrt{3}}{3}$, the arc tan of which is $\pi/6$.

Suppose next, w is the angle between two lines $y = \sqrt{3}x - 1$ and $y = -\sqrt{3}x + 9$.

Then, $\tan \pi/3 = \sqrt{3} = a$, and $\tan 2\pi/3 = \tan(\pi - \pi/3) = -\tan \pi/3 = -\sqrt{3} = c$.

That is, $u = \pi/3$, and $v = 2\pi/3$. So $w = \pi/3$.

And putting the two lines in the x-y plane, we can put them the way below:

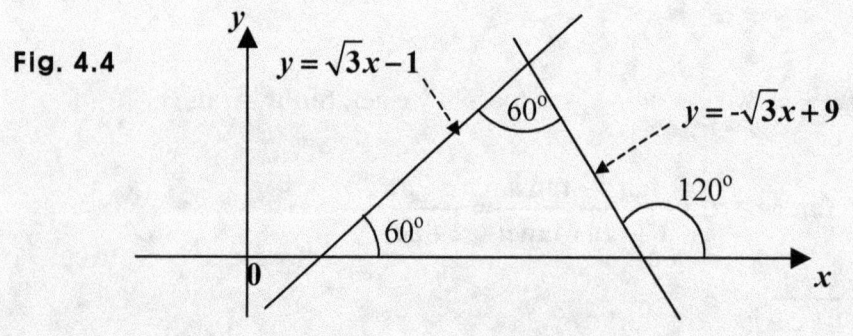

Fig. 4.4

Then, we get: $\dfrac{c-a}{1+ca} = \dfrac{-\sqrt{3}-\sqrt{3}}{1+(-\sqrt{3})\cdot\sqrt{3}} = \dfrac{-2\sqrt{3}}{1-3} = \sqrt{3}$, the arc tan of which is $\pi/3$.

Suppose next, *w* is the angle between two lines $y = \frac{\sqrt{3}}{3}x - 1$ and $y = -\sqrt{3}x + 9$.

Then, **tan $\pi/6 = \frac{\sqrt{3}}{3} = a$, and tan $2\pi/3$ = tan($\pi - \pi/3$) = -tan $\pi/3$ = $-\sqrt{3} = c$.**

That is, *u = π/6*, and *v = 2π/3*.   So *w = π/2*.

And putting the two lines in the *x-y* plane, we can put them the way below:

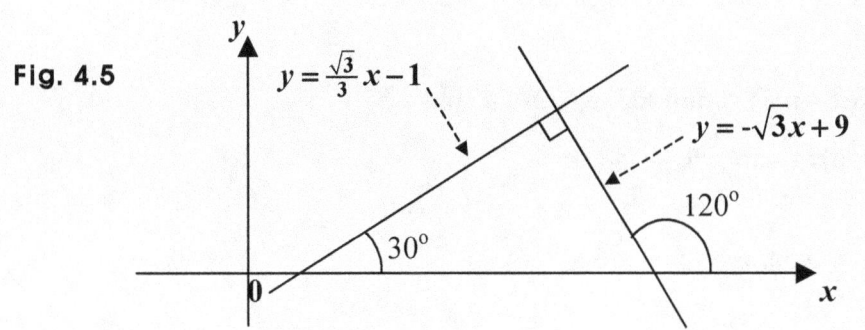

Fig. 4.5

Then, we get: $\dfrac{c-a}{1+ca} = \dfrac{-\sqrt{3}-\frac{\sqrt{3}}{3}}{1+(-\sqrt{3})\cdot\frac{\sqrt{3}}{3}} = \dfrac{\frac{-3\sqrt{3}-\sqrt{3}}{3}}{0}$, which means *w = π/2*.

Suppose next, *w* is the angle between two lines $y = \frac{\sqrt{3}}{3}x - 1$ and $y = -\frac{\sqrt{3}}{3}x + 9$.

Then, **tan $\pi/6 = \frac{\sqrt{3}}{3} = a$, and tan $5\pi/6$ = tan($\pi - \pi/6$) = -tan $\pi/6$ = $-\frac{\sqrt{3}}{3} = c$.**

That is, *u = π/6*, and *v = 5π/6*.   So *w = 2π/3*.

And putting the two lines in the *x-y* plane, we can put them the way below:

Fig. 4.6

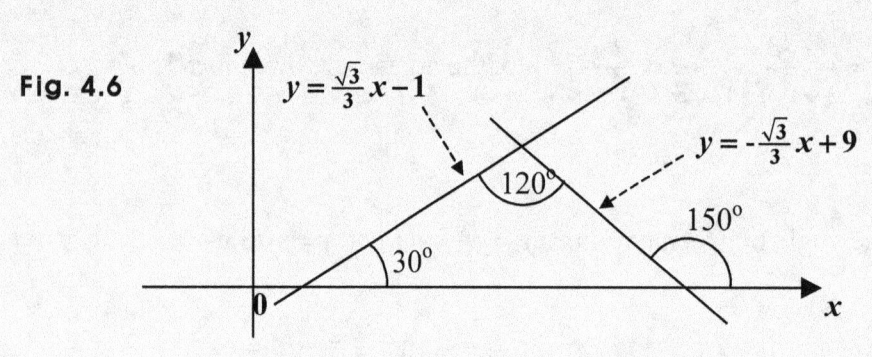

Then, we get: $\dfrac{c-a}{1+ca} = \dfrac{-\frac{\sqrt{3}}{3}-\frac{\sqrt{3}}{3}}{1+(-\frac{\sqrt{3}}{3})\cdot\frac{\sqrt{3}}{3}} = \dfrac{-\frac{2\sqrt{3}}{3}}{1-\frac{1}{3}} = \dfrac{-\frac{2\sqrt{3}}{3}}{\frac{2}{3}} = -\dfrac{2\sqrt{3}}{2} = -\sqrt{3}$, the arc tan of which

is **2π/3**.

Suppose next, **w** is the angle between two lines $y = -\sqrt{3}x + 5$ and $y = -\frac{\sqrt{3}}{3}x + 4$.

Then, **tan 2π/3 = tan(π − π/3) = -tan π/3 = -$\sqrt{3}$ = a**, and

tan 5π/6 = tan (π − π/6) = -tan π/6 = -$\frac{\sqrt{3}}{3}$ = c.

That is, **u = 2π/3**, and **v = 5π/6**. So **w = π/6**.

And putting the two lines in the **x-y** plane, we can put them the way below:

Fig. 4.7

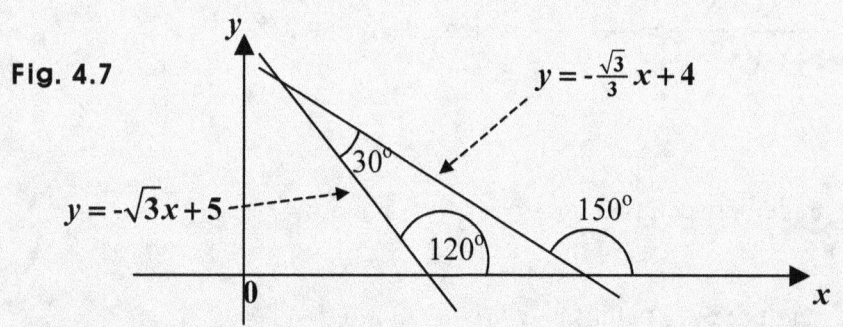

Then, we get: $\dfrac{c-a}{1+ca} = \dfrac{-\frac{\sqrt{3}}{3}-(-\sqrt{3})}{1+(-\frac{\sqrt{3}}{3})\cdot(-\sqrt{3})} = \dfrac{-\frac{\sqrt{3}}{3}+\sqrt{3}}{1+\frac{\sqrt{3}}{3}\cdot\sqrt{3}} = \dfrac{\frac{-\sqrt{3}+3\sqrt{3}}{3}}{2} = \dfrac{2\sqrt{3}}{6} = \dfrac{\sqrt{3}}{3}$, the arc tan

of which is **π/6**.